THE HIDDEN HEALERS: DECODING THE SECRETOME OF MESENCHYMAL STEM CELLS

DR. JAMES UTLEY PHD
DR. DANIEL BRIGGS DRPH

ABOUT THE AUTHOR

Meet Dr. James Utley, PhD. He's not your typical Immunohematology guru. For over two decades, he's been crafting a niche in cellular therapy across the U.S. A proud alum of Johns Hopkins, he took a detour to the Department of Defense, where he reimagined the whole cellular transfusion scene.

But here's where it gets intriguing: James is a Biohacker at heart. He's been on the cutting edge, self-experimenting and pushing the boundaries of CRISPR and genetic engineering like a true avant-garde scientist.

Then there's his stint at Banner Health. Not just any role, mind you. He was the Technical Director of all the blood banks and transfusion services, overseeing a monumental 150K successful cellular transfusions with his artisanal touch. But James isn't just about numbers. He's penned thoughts in some of the most avant-garde medical journals and has been the force behind some truly innovative FDA-approved breakthroughs.

And here's the kicker: this former Navy Scientist traded in his lab coat to chart the unexplored waters of the stem cell revolution. Some even call him the "pirate" of the cellular world. Now, as the Chief Scientific Officer at Auragens, he's making waves and a difference in the world every single day. Cheers to the unconventional trailblazer and world-changer!

www.auragens.com | Instagram: @theauragens | @dr.james.utley

ABOUT THE CO-AUTHOR

With over twenty years of experience as a realistic visionary and a beneficial entrepreneur, Dr. Dan Briggs is the Founder, President, and CEO of multiple successful companies.

In the medical field Dan is the Founder, President & CEO of Auragens, the premier stem cell and research center in the world and located in Panama City, Panama. Spanning the entire 48th floor of the Oceania Business Tower, Auragens has created the world's leading research facility and has attracted top PhD's and medical doctors from the Americas and Europe in its pursuit to improve the standard of care and treatment. He formed a 501c3 nonprofit, primary care company, The Neighborhood Clinic, where he also serves as Chairman and CEO. With multiple locations The Neighborhood Clinic's doctors, and nurse practitioners, can see up to 500 underserved and rural patients per day that previously had no access to medical care. Dan also founded MDX Labs Inc, a CLIA and COLA Certified, high complexity lab network headquartered in Henderson, Nevada. A frontrunner and innovator in the molecular & clinical diagnostics space,

MDX was recognized as "Top Small Business" and "Best Workplace" in Nevada. Additionally, and to further his passion in supporting the community Dan founded and serves as Chairman of FOUNDA-TIONS, a 501c3 charitable organization, that provides support and donations to charities globally assisting those in need and breaking down barriers in access to healthcare. In 2023, Dr. Dan Briggs was named Healthcare Executive of the Year by Nevada Business Magazine and a jury of his peers.

Along with his role in his companies, Dan also has served, and continues to serve, on several boards where he acts as advisor, trustee, and mentor. These include NeoVolta (NASDAQ: NEOV), University of Northwestern Ohio (UNOH), Las Vegas HEALS, and Big Brothers Big Sisters.

Dan's career has firm roots in public health, policy, and advocacy, long-held personal passions of his. Dan was a member of the advance team of President George W. Bush during the first US and Russia Summit with Vladimir Putin, held in St. Petersburg, Russia. Dan then oversaw campaigns on behalf of the Governor of California for the Office of Members Services. After relocating to Nevada, Dan was a Founding Member of the Las Vegas World Affairs Council -- a bipartisan organization dedicated to engaging and educating Americans on international affairs and foreign policy. His desire for public service resulted in him being recruited to run as a candidate for the Nevada State Assembly District 20 in 2008. He lost.

Dan earned his Doctorate Degrees in Public Health (DrPH) and Doctor of Laws (LLD) from University of Northwestern Ohio, studied law at Thomas Jefferson School of Law in San Diego, earned his master's degree in Russian, East European, and Eurasian Studies at Stanford University, and an undergraduate degree in Political Science from Pepperdine University. He is married with three children and splits his time between the USA and Panama.

1st edition 2023

DISCLAIMER

The content provided in this book is for informational and educational purposes only and is not intended as, nor should it be considered a substitute for, professional medical advice. Do not use the information in this book for diagnosing or treating any medical or health condition. If you have or suspect you have a medical problem, promptly consult your professional healthcare provider. Always seek the advice of your physician or other qualified health provider with any questions you may have regarding a medical condition. Never disregard professional medical advice or delay in seeking it because of something you have read in this book.

CONTENTS

CHAPTER ONE

INTRODUCTION TO MESENCHYMAL STEM CELLS

Overview of the Secretome of Mesenchymal Stem Cells

Mesenchymal stem cells (MSCs) have gained significant attention in the field of regenerative medicine due to their remarkable regenerative properties. While the ability of MSCs to differentiate into various cell types has been extensively studied, recent research has revealed that their therapeutic effects are not solely attributed to their differentiation potential. Instead, it has been discovered that MSCs secrete a wide range of bioactive molecules, collectively known as the secretome, which play a crucial role in tissue repair and regeneration (Joshi, Dehghan Abnavi, & Kothapalli, 2019). This section will provide an overview of the secretome of MSCs and its role in various regenerative processes.

The secretome of MSCs consists of a complex mixture of growth factors, cytokines, and chemokines that exert diverse effects on the surrounding cells and tissues. These bioactive molecules are released by MSCs in response to specific cues from the microenvironment, such as inflammation or injury. The secretome acts in a paracrine manner, meaning that it acts on nearby cells rather than directly differentiating into those cell types. This paracrine signaling is a key mechanism through which MSCs exert their regenerative effects (Deschepper, Paquet, & Petite, 2014).

Applications of Mesenchymal Stem Cells

Mesenchymal stem cells (MSCs) have emerged as pivotal entities within the domain of regenerative medicine, attributed to their profound regenerative capabilities. These multipotent cells exhibit the capacity for differentiation into diverse cellular phenotypes and are characterized by the secretion of an extensive array of bioactive molecules, collectively termed the secretome. The secretome of MSCs is instrumental in orchestrating the therapeutic efficacy of these cells in tissue regeneration and repair (Lai et al., 2019). This discourse aims to delineate the utilization of MSCs and their secretome across various spectrums of regenerative medicine, underscoring their potential in advancing therapeutic interventions.

Multipotency

One of the key properties of MSCs is their ability to differentiate into multiple cell types. MSCs have the capacity to undergo numerous cell divisions while maintaining their undifferentiated state, ensuring a constant supply of cells for tissue repair and regeneration. Additionally, MSCs possess the potential to differentiate into various cell lineages, including osteocytes (bone cells), chondrocytes (cartilage cells), adipocytes (fat cells), and myocytes (muscle cells). This multipotency makes MSCs a versatile tool for tissue engineering and regenerative medicine (Bellei, Migliano, & Picardo, 2022).

Immunomodulatory Effects

MSCs have been shown to possess potent immunomodulatory properties, meaning they can regulate and modulate the immune response. These cells can suppress the activation and proliferation of immune cells, such as T cells, B cells, and natural killer cells, while promoting the expansion of regulatory T cells. This immunomodulatory effect is mediated by the secretion of various factors, including growth factors, cytokines, and chemokines, collectively known as the secretome of MSCs (Joshi, Dehghan Abnavi, & Kothapalli, 2019).

Inflammation is a natural response to tissue injury or infection; however, chronic inflammation can impede the healing process and lead to tissue damage. MSCs have been shown to possess potent anti-inflammatory effects through the secretion of anti-inflammatory cytokines, such as interleukin-10 (IL-10) and transforming growth factor-beta (TGF-β). These cytokines help to suppress the inflammatory response and promote the resolution of inflammation, creating a favorable environment for tissue regeneration (Lai et al., 2019).

Inflammation is a common response to tissue injury and can impede the healing process. The secretome of MSCs possesses potent anti-inflammatory effects. Several molecules present in the secretome contribute to the suppression of inflammation. Some of the key molecules involved in the anti-inflammatory effects include:

Growth Factors: Transforming growth factor-beta (TGF-β) and hepatocyte growth factor (HGF).

Cytokines: Interleukin-10 (IL-10) and interleukin-1 receptor antagonist (IL-1ra).

Chemokines: Macrophage inflammatory protein-1 alpha (MIP-1α) and monocyte chemoattractant protein-1 (MCP-1).

These molecules inhibit the production of pro-inflammatory cytokines, reduce immune cell infiltration, and promote the resolution of inflammation, thereby facilitating tissue healing (Bellei, Migliano, & Picardo, 2022).

Angiogenic Potential

Angiogenesis, the formation of new blood vessels, is a crucial process in tissue repair and regeneration. MSCs have been found to promote angiogenesis through the secretion of specific factors that stimulate the growth and migration of endothelial cells, the building blocks of blood vessels. Some of the growth factors involved in angiogenesis include vascular endothelial growth factor (VEGF), fibroblast growth factor (FGF), and platelet-derived growth factor (PDGF). Additionally, MSCs secrete cytokines and chemokines that attract endothelial cells to the site of injury and facilitate the formation of new blood vessels (Joshi, Dehghan Abnavi, & Kothapalli, 2019).

Angiogenesis, the formation of new blood vessels, is a critical process in tissue repair and regeneration. The secretome of MSCs contains several growth factors, cytokines, and chemokines that promote angiogenesis. Some of the key molecules involved in angiogenesis include:

Growth Factors: Vascular endothelial growth factor (VEGF), fibroblast growth factor (FGF), platelet-derived growth factor (PDGF), hepatocyte growth factor (HGF), and angiopoietin-1 (Ang-1).

Cytokines: Interleukin-8 (IL-8), interleukin-6 (IL-6), and transforming growth factor-beta (TGF-β).

Chemokines: Stromal cell-derived factor-1 (SDF-1) and monocyte chemoattractant protein-1 (MCP-1).

The secretion of these molecules by MSCs promotes the recruitment and proliferation of endothelial cells, leading to the formation of new blood vessels and improved blood supply to the damaged tissues (Lai et al., 2019).

Scar Reduction

Scar formation is a natural part of the wound healing process; however, excessive scarring can lead to functional and aesthetic complications. MSCs have been shown to have a significant impact on scar reduction through their secretome. The growth factors released by MSCs, such as transforming growth factor-beta (TGF-β) and epidermal growth factor (EGF), play a crucial role in regulating the synthesis and remodeling of the extracellular matrix, which is responsible for scar formation. Additionally, MSCs secrete cytokines and chemokines that modulate the inflammatory response and promote tissue regeneration, leading to reduced scar formation (Bellei, Migliano, & Picardo, 2022).

Scar formation is a common consequence of tissue injury and can impair tissue function and aesthetics. The secretome of MSCs has been found to possess scar-reducing properties. Several growth factors, cytokines, and chemokines present in the secretome contribute to scar reduction. Some of the key molecules involved in scar reduction include:

Growth Factors: Transforming growth factor-beta (TGF-β), epidermal growth factor (EGF), and insulin-like growth factor-1 (IGF-1).

Cytokines: Interleukin-10 (IL-10) and tumor necrosis factor-alpha (TNF-α).

Chemokines: Regulated on activation, normal T cell expressed and secreted (RANTES) and macrophage inflammatory protein-1 alpha (MIP-1α). These molecules modulate the inflammatory response, promote tissue remodeling, and inhibit excessive collagen deposition, thereby reducing scar formation and promoting tissue regeneration (Joshi, Dehghan Abnavi, & Kothapalli, 2019).

Cell Proliferation

MSCs have the ability to stimulate cell proliferation, which is essential for tissue regeneration and repair. The secretome of MSCs contains growth factors, such as insulin-like growth factor (IGF) and hepatocyte growth factor (HGF), which promote cell division and proliferation. These growth factors act on various cell types, including fibroblasts, endothelial cells, and epithelial cells, to enhance their proliferation and facilitate tissue regeneration (Lai et al., 2019).

Enhancing cell proliferation is crucial for tissue regeneration and repair. The secretome of MSCs contains various factors that stimulate cell proliferation. Some of the key molecules involved in cell proliferation include:

Growth Factors: Epidermal growth factor (EGF), fibroblast growth factor (FGF), insulin-like growth factor-1 (IGF-1), and platelet-derived growth factor (PDGF).

Cytokines: Interleukin-1 alpha (IL-1α) and interleukin-6 (IL-6).

Chemokines: Growth-regulated oncogene-alpha (GRO-α) and macrophage inflammatory protein-1 beta (MIP-1β).

These molecules promote the proliferation of various cell types, including fibroblasts, endothelial cells, and epithelial cells, facilitating tissue regeneration and repair (Bellei, Migliano, & Picardo, 2022).

Anti-microbial Effects

In addition to their regenerative properties, MSCs have been found to exhibit antimicrobial effects. The secretome of MSCs contains antimicrobial peptides and proteins that can directly inhibit the growth and proliferation of bacteria, viruses, and fungi. These antimicrobial factors, such as cathelicidin and defensins, help to protect the injured tissue from infection and promote a sterile environment for tissue regeneration (Joshi, Dehghan Abnavi, & Kothapalli, 2019).

In addition to their regenerative properties, MSCs and their secretome exhibit antimicrobial effects. The secretome contains molecules that possess antimicrobial properties and can inhibit the growth of various pathogens.

Some of the key molecules involved in the anti-microbial effects include:

Growth Factors: Hepatocyte growth factor (HGF) and granulocyte-macrophage colony-stimulating factor (GM-CSF).

Cytokines: Interleukin-1 beta (IL-1β) and interferon-gamma (IFN-γ).

Chemokines: Regulated on activation, normal T cell expressed and secreted (RANTES) and macrophage inflammatory protein-1 alpha (MIP-1α).

These molecules enhance the immune response against pathogens and promote the clearance of microbial infections (Lai et al., 2019).

Anti-Apoptotic Effects

Apoptosis, or programmed cell death, is a natural process that occurs during tissue development and homeostasis. However, excessive cell death can hinder tissue regeneration. MSCs have been shown to have anti-apoptotic effects through the secretion of factors that inhibit cell death pathways and promote cell survival. Some of the growth factors involved in anti-apoptotic effects include insulin-like growth factor-1 (IGF-1) and hepatocyte growth factor (HGF). These factors help to protect cells from apoptosis and enhance tissue regeneration (Bellei, Migliano, & Picardo, 2022).

Apoptosis, or programmed cell death, can occur in injured tissues and hinder the regeneration process. The secretome of MSCs contains factors that exhibit anti-apoptotic effects, promoting cell survival and tissue regeneration. Some of the key molecules involved in the anti-apoptotic effects include:

Growth Factors: Insulin-like growth factor-1 (IGF-1) and nerve growth factor (NGF).

Cytokines: Interleukin-6 (IL-6) and tumor necrosis factor-alpha (TNF-α).

Chemokines: Stromal cell-derived factor-1 (SDF-1) and monocyte chemoattractant protein-1 (MCP-1).

These molecules inhibit apoptotic pathways, promote cell survival, and enhance tissue regeneration (Joshi, Dehghan Abnavi, & Kothapalli, 2019).

In summary, the secretome of MSCs exhibits a wide range of applications in regenerative medicine. The growth factors, cytokines, and chemokines present in the secretome contribute to angiogenesis, scar reduction, cell proliferation, immunomodulation, anti-inflammatory effects, anti-microbial effects, and anti-apoptotic effects. Understanding the therapeutic potential of the secretome is crucial for harnessing the regenerative properties of MSCs and developing novel therapeutic strategies for various diseases and injuries.

CHAPTER TWO

ANGIOGENESIS AND THE SECRETOME

Introduction to Angiogenesis

Angiogenesis, the formation of new blood vessels from pre-existing ones, plays a crucial role in various physiological and pathological processes, including wound healing, tissue repair, and tumor growth. The process of angiogenesis involves the migration, proliferation, and differentiation of endothelial cells, which line the inner surface of blood vessels. Mesenchymal stem cells (MSCs) have emerged as a promising therapeutic tool for promoting angiogenesis due to their ability to secrete a wide range of bioactive molecules, collectively known as the secretome (Durum et al., n.d.).

The secretome of MSCs contains a diverse array of growth factors, cytokines, and chemokines that contribute to the pro-angiogenic effects of these cells. These bioactive molecules act in a coordinated manner to stimulate endothelial cell proliferation, migration, and tube formation, ultimately leading to the formation of new blood vessels (Durum et al., n.d.). In this section, we will explore the key growth factors, cytokines, and chemokines involved in angiogenesis and their role in the secretome of MSCs.

Growth Factors involved in Angiogenesis

Growth factors are potent signaling molecules that regulate various cellular processes, including angiogenesis. Within the secretome of MSCs, several growth factors have been identified to play a crucial role in promoting angiogenesis. Some of the key growth factors involved in angiogenesis include:

- Vascular Endothelial Growth Factor (VEGF): VEGF is one of the most well-known and potent pro-angiogenic factors. It stimulates endothelial cell proliferation, migration, and tube formation, and also promotes the recruitment of endothelial progenitor cells to the site of angiogenesis (Ito et al., 2013).

- Fibroblast Growth Factor (FGF): FGFs are a family of growth factors that play a crucial role in angiogenesis. They stimulate endothelial cell proliferation, migration, and tube formation, and also promote the production of other pro-angiogenic factors (Ito et al., 2013).

- Platelet-Derived Growth Factor (PDGF): PDGF is a potent mitogen for endothelial cells and plays a crucial role in angiogenesis. It stimulates endothelial cell proliferation and migration, and also promotes the recruitment of pericytes, which stabilize newly formed blood vessels (Ravindran et al., 2006).

- Angiopoietins: Angiopoietins are a family of growth factors that regulate angiogenesis by interacting with endothelial cells and promoting their survival, proliferation, and migration. Angiopoietin-1 and Angiopoietin-2 are the two most well-studied members of this family (Durum et al., n.d.).

Cytokines involved in Angiogenesis

Cytokines are small proteins that regulate immune responses and play a crucial role in angiogenesis. Within the secretome of MSCs, several cytokines have been identified to have pro-angiogenic effects. Some of the key cytokines involved in angiogenesis include:

- Interleukin-8 (IL-8): IL-8 is a potent pro-angiogenic cytokine that stimulates endothelial cell migration and tube formation. It also promotes the recruitment of endothelial progenitor cells to the site of angiogenesis (Ito et al., 2013).

- Interleukin-6 (IL-6): IL-6 is a multifunctional cytokine that plays a crucial role in angiogenesis. It stimulates endothelial cell proliferation and migration and also promotes the production of other pro-angiogenic factors (Ravindran et al., 2006).

- Tumor Necrosis Factor-alpha (TNF-alpha): TNF-alpha is a pro-inflammatory cytokine that also plays a role in angiogenesis. It stimulates endothelial cell proliferation and migration and promotes the production of other pro-angiogenic factors (Ito et al., 2013).

- Interferon-gamma (IFN-gamma): IFN-gamma is a cytokine that regulates immune responses and also has pro-angiogenic effects. It stimulates endothelial cell proliferation and migration and promotes the production of other pro-angiogenic factors (Ravindran et al., 2006).

Chemokines involved in Angiogenesis

Chemokines are a family of small proteins that regulate the migration and activation of immune cells. Within the secretome of MSCs, several chemokines have been identified to have pro-angiogenic effects. Some of the key chemokines involved in angiogenesis include:

- Stromal cell-derived factor-1 (SDF-1): SDF-1 is a chemokine that plays a crucial role in angiogenesis. It stimulates endothelial cell migration and promotes the recruitment of endothelial progenitor cells to the site of angiogenesis (Ito et al., 2013).

- Monocyte Chemoattractant Protein-1 (MCP-1): MCP-1 is a chemokine that promotes the recruitment of monocytes and macrophages to the site of angiogenesis. These immune cells play a crucial role in the angiogenic process (Ravindran et al., 2006).

- Interleukin-8 (IL-8): IL-8, in addition to its role as a cytokine, also acts as a chemokine and promotes the migration of endothelial cells to the site of angiogenesis (Ito et al., 2013).

- Growth-Regulated Oncogene-alpha (GRO-alpha): GRO-alpha is a chemokine that stimulates endothelial cell migration and promotes the formation of new blood vessels (Ravindran et al., 2006).

In conclusion, angiogenesis is a complex process that involves the coordinated action of various growth factors, cytokines, and chemokines. The secretome of MSCs contains a rich repertoire of these bioactive molecules, which contribute to the pro-angiogenic effects of these cells. Understanding the role of the secretome in angiogenesis is crucial for harnessing the regenerative potential of MSCs in various therapeutic applications.

Further Insights into Growth Factors Contributing to Angiogenesis

Angiogenesis, the formation of new blood vessels from pre-existing ones, is a crucial process in tissue repair and regeneration. Mesenchymal stem cells (MSCs) have been shown to play a significant role in promoting angiogenesis through the secretion of various growth factors, cytokines, and chemokines (Kumar, Kumar, Singh, & Goel, n.d.). These factors act in a coordinated manner to stimulate the proliferation, migration, and differentiation of endothelial cells, leading to the formation of new blood vessels.

Vascular Endothelial Growth Factor (VEGF)

Vascular endothelial growth factor (VEGF) is one of the most potent angiogenic factors secreted by MSCs. It promotes angiogenesis by binding to its receptors on endothelial cells, stimulating their proliferation and migration. VEGF also enhances the permeability of blood vessels, allowing the recruitment of immune cells and the delivery of nutrients and oxygen to the newly formed vessels (Adas et al., 2016).

Fibroblast Growth Factor (FGF)

Fibroblast growth factors (FGFs) are a family of growth factors that play a crucial role in angiogenesis. MSCs secrete various FGFs, including FGF-2 and FGF-7, which promote the proliferation and migration of endothelial cells. FGFs also stimulate the production of other angiogenic factors, such as VEGF, further enhancing the angiogenic response (Wang et al., 2020).

Platelet-Derived Growth Factor (PDGF)

Platelet-derived growth factor (PDGF) is another growth factor secreted by MSCs that contributes to angiogenesis. PDGF stimulates the proliferation and migration of endothelial cells, as well as the recruitment of pericytes, which stabilize the newly formed blood vessels. PDGF also plays a role in the recruitment of MSCs to the site of injury, further enhancing the angiogenic response (Konala, Bhonde, & Pal, 2020).

Transforming Growth Factor-Beta (TGF-β)

Transforming growth factor-beta (TGF-β) is a multifunctional cytokine that regulates various cellular processes, including angiogenesis. MSCs secrete TGF-β, which can promote or inhibit angiogenesis depending on the context. In certain conditions, TGF-β stimulates the production of other angiogenic factors, such as VEGF, promoting angiogenesis. However, in other situations, TGF-β can inhibit angiogenesis by suppressing the proliferation and migration of endothelial cells (Wang et al., 2020).

Angiopoietins

Angiopoietins are a family of growth factors that play a crucial role in angiogenesis. MSCs secrete angiopoietin-1 (Ang-1) and angiopoietin-2 (Ang-2), which bind to their receptors on endothelial cells, promoting vessel stabilization and remodeling. Ang-1 promotes the recruitment of pericytes and smooth muscle cells, while Ang-2 destabilizes blood vessels, allowing for vessel sprouting and remodeling (Konala, Bhonde, & Pal, 2020).

Insulin-like Growth Factor (IGF)

Insulin-like growth factor (IGF) is a growth factor that regulates cell growth and differentiation. MSCs secrete IGF, which promotes angiogenesis by stimulating the proliferation and migration of endothelial cells. IGF also enhances the survival of endothelial cells and promotes the formation of capillary-like structures (Adas et al., 2016).

Hepatocyte Growth Factor (HGF)

Hepatocyte growth factor (HGF) is a growth factor secreted by MSCs that plays a crucial role in angiogenesis. HGF stimulates the proliferation and migration of endothelial cells, promoting the formation of new blood vessels. HGF also enhances the production of other angiogenic factors, such as VEGF, further enhancing the angiogenic response (Adas et al., 2016).

In summary, MSCs secrete a variety of growth factors that contribute to angiogenesis. These growth factors act in a coordinated manner to stimulate the proliferation, migration, and differentiation of endothelial cells, leading to the formation of new blood vessels. The secretion of these growth factors by MSCs highlights their potential as a therapeutic tool for promoting angiogenesis and enhancing tissue repair and regeneration (Kumar, Kumar, Singh, & Goel, n.d.).

Further insights into Cytokines contributing to Angiogenesis

Angiogenesis, the formation of new blood vessels from pre-existing ones, is a crucial process in tissue repair and regeneration. Mesenchymal stem cells (MSCs) have been shown to play a significant role in promoting angiogenesis through the secretion of various factors, including cytokines (Konala, Bhonde, & Pal, 2020). In this section, we will explore the cytokines involved in angiogenesis and their role in the healing process.

Vascular Endothelial Growth Factor (VEGF)

Vascular endothelial growth factor (VEGF) is one of the key cytokines involved in angiogenesis. It promotes the growth and migration of endothelial cells, which are essential for the formation of new blood vessels. MSCs have been found to secrete VEGF, which stimulates the proliferation and migration of endothelial cells, leading to the formation of new blood vessels. This cytokine also enhances the permeability of blood vessels, allowing for the efficient delivery of nutrients and oxygen to the healing tissues (Chernoff, n.d.).

Fibroblast Growth Factor (FGF)

Fibroblast growth factor (FGF) is another important cytokine involved in angiogenesis. It stimulates the proliferation and migration of endothelial cells, promoting the formation of new blood vessels. MSCs have been shown to secrete various isoforms of FGF, such as FGF-2 and FGF-7, which contribute to angiogenesis. FGF also plays a role in the recruitment of other cells involved in tissue repair, such as fibroblasts and smooth muscle cells (Khan, Barry, O'Brien, & Kerin, n.d.).

Platelet-Derived Growth Factor (PDGF)

Platelet-derived growth factor (PDGF) is a cytokine that plays a crucial role in angiogenesis by stimulating the proliferation and migration of endothelial cells. MSCs have been found to secrete PDGF, which promotes the formation of new blood vessels. PDGF also stimulates the recruitment of pericytes, which are important for stabilizing the newly formed blood vessels and promoting their maturation (Kim, Liu, Kucia, & Ratajczak, 2011).

Transforming Growth Factor-Beta (TGF-β)

Transforming growth factor-beta (TGF-β) is a multifunctional cytokine involved in various cellular processes, including angiogenesis. MSCs secrete TGF-β, which can stimulate the proliferation and migration of endothelial cells. However, the role of TGF-β in angiogenesis is complex, as it can also have inhibitory effects on blood vessel formation under certain conditions. The balance between the pro-angiogenic and anti-angiogenic effects of TGF-β is crucial for proper tissue healing and regeneration (Konala, Bhonde, & Pal, 2020).

Interleukins (ILs)

Interleukins (ILs) are a group of cytokines that play diverse roles in the immune system and tissue repair processes. Several ILs have been implicated in angiogenesis. For example, IL-8 is known to promote the migration and proliferation of endothelial cells, contributing to the formation of new blood vessels. IL-6 and IL-1β have also been shown to have pro-angiogenic effects. MSCs can secrete these ILs, thereby enhancing angiogenesis and facilitating tissue repair (Chernoff, n.d.).

Other Cytokines

In addition to the cytokines mentioned above, MSCs secrete various other cytokines that can influence angiogenesis. These include hepatocyte growth factor (HGF), granulocyte colony-stimulating factor (G-CSF), and stromal cell-derived factor-1 (SDF-1). These cytokines can promote the migration, proliferation, and survival of endothelial cells, contributing to the formation of new blood vessels (Kim, Liu, Kucia, & Ratajczak, 2011).

The secretion of these cytokines by MSCs creates a favorable microenvironment for angiogenesis, facilitating the repair and regeneration of damaged tissues. The coordinated action of these cytokines promotes the growth and remodeling of blood vessels, ensuring an adequate blood supply to the healing tissues. Understanding the role of these cytokines in angiogenesis is crucial for harnessing the regenerative potential of MSCs and developing effective therapeutic strategies for various diseases and injuries (Konala, Bhonde, & Pal, 2020).

Further insights into Chemokines contributing to Angiogenesis

Angiogenesis, the formation of new blood vessels from pre-existing ones, is a crucial process in tissue repair and regeneration. Mesenchymal stem cells (MSCs) have been shown to play a significant role in promoting angiogenesis through the secretion of various factors, including growth factors, cytokines, and chemokines (Sun et al., 2005). In this section, we will explore the chemokines involved in angiogenesis and their role in facilitating the formation of new blood vessels.

CXCL12 (SDF-1)

One of the key chemokines involved in angiogenesis is CXCL12, also known as stromal cell-derived factor 1 (SDF-1). CXCL12 is secreted by MSCs and acts as a chemoattractant for endothelial progenitor cells (EPCs) and endothelial cells. It promotes the migration and recruitment of these cells to the site of injury, where they contribute to the formation of new blood vessels. CXCL12 also enhances the survival and proliferation of endothelial cells, further supporting angiogenesis (Sun et al., 2005).

CXCL8 (IL-8)

CXCL8, also known as interleukin-8 (IL-8), is another chemokine secreted by MSCs that plays a role in angiogenesis. It acts as a potent chemoattractant for endothelial cells and promotes their migration towards the site of injury. CXCL8 also stimulates the proliferation and tube formation of endothelial cells, facilitating the formation of new blood vessels. Additionally, CXCL8 has been shown to enhance the production of other angiogenic factors, such as vascular endothelial growth factor (VEGF), further promoting angiogenesis (Arsentieva et al., 2020).

CCL2 (MCP-1)

CCL2, also known as monocyte chemoattractant protein-1 (MCP-1), is a chemokine secreted by MSCs that contributes to angiogenesis. It acts as a chemoattractant for monocytes and macrophages, which play a crucial role in the angiogenic process. CCL2 promotes the recruitment of these immune cells to the site of injury, where they release pro-angiogenic factors and contribute to the remodeling of blood vessels. Additionally, CCL2 has been shown to enhance the migration and tube formation of endothelial cells, further supporting angiogenesis (Arsentieva et al., 2020).

CXCL1 (GRO-α)

CXCL1, also known as growth-regulated oncogene alpha (GRO-α), is a chemokine secreted by MSCs that has angiogenic properties. It acts as a chemoattractant for endothelial cells and promotes their migration towards the site of injury. CXCL1 also stimulates the proliferation and tube formation of endothelial cells, facilitating the formation of new blood vessels. Additionally, CXCL1 has been shown to enhance the production of other angiogenic factors, such as VEGF, further promoting angiogenesis (Sun et al., 2005).

CXCL5 (ENA-78)

CXCL5, also known as epithelial-derived neutrophil-activating peptide 78 (ENA-78), is a chemokine secreted by MSCs that contributes to angiogenesis. It acts as a chemoattractant for endothelial cells and promotes their migration towards the site of injury. CXCL5 also stimulates the proliferation and tube formation of endothelial cells, facilitating the formation of new blood vessels. Additionally, CXCL5 has been shown to enhance the production of other angiogenic factors, such as VEGF, further promoting angiogenesis (Arsentieva et al., 2020).

CXCL6 (GCP-2)

CXCL6, also known as granulocyte chemotactic protein-2 (GCP-2), is a chemokine secreted by MSCs that plays a role in angiogenesis. It acts as a chemoattractant for endothelial cells and promotes their migration towards the site of injury. CXCL6 also stimulates the proliferation and tube formation of endothelial cells, facilitating the formation of new blood vessels. Additionally, CXCL6 has been shown to enhance the production of other angiogenic factors, such as VEGF, further promoting angiogenesis (Sun et al., 2005).

In summary, MSCs secrete various chemokines that contribute to angiogenesis, the formation of new blood vessels. Chemokines such as CXCL12, CXCL8, CCL2, CXCL1, CXCL5, and CXCL6 play crucial roles in promoting the migration, recruitment, proliferation, and tube formation of endothelial cells, facilitating the process of angiogenesis. Understanding the role of these chemokines in angiogenesis can provide valuable insights into the regenerative potential of MSCs and their therapeutic applications in promoting tissue repair and regeneration (Sun et al., 2005; Arsentieva et al., 2020).

CHAPTER THREE

SCAR REDUCTION AND THE SECRETOME

Introduction to Scar Reduction

Scarring is a natural part of the healing process that occurs after an injury or surgery. While scars serve as a protective barrier, they can also be aesthetically displeasing and cause functional limitations. Scar reduction is a crucial aspect of regenerative medicine, and mesenchymal stem cells (MSCs) have shown promising potential in this field (Deszcz, 2023).

The Role of MSCs in Scar Reduction

Mesenchymal stem cells have gained significant attention in the field of scar reduction due to their regenerative properties. These cells have the ability to differentiate into various cell types, including fibroblasts, which are responsible for producing the extracellular matrix (ECM) that forms scars. By modulating the production and remodeling of the ECM, MSCs can play a vital role in reducing scar formation (Costela-Ruiz et al., 2022).

Growth Factors Involved in Scar Reduction

The secretome of MSCs contains a variety of growth factors that contribute to scar reduction. These growth factors promote tissue regeneration, modulate the inflammatory response, and regulate the synthesis and degradation of the ECM. Some of the key growth factors involved in scar reduction include:

- Transforming Growth Factor-beta (TGF-β): TGF-β is a multifunctional growth factor that plays a crucial role in wound healing and scar formation. In scar reduction, TGF-β promotes the synthesis of collagen and other ECM components, leading to scar remodeling and improved tissue architecture (Deszcz, 2023).

- Platelet-Derived Growth Factor (PDGF): PDGF is involved in various cellular processes, including cell proliferation, migration, and angiogenesis. In scar reduction, PDGF stimulates the migration of fibroblasts to the wound site, promoting the formation of new tissue and reducing scar formation (Costela-Ruiz et al., 2022).

- Epidermal Growth Factor (EGF): EGF is known for its role in promoting cell proliferation and migration. In scar reduction, EGF accelerates the healing process by stimulating the growth and migration of epithelial cells, leading to improved wound closure and reduced scar formation (Costela-Ruiz et al., 2022).

Cytokines Involved in Scar Reduction

In addition to growth factors, the secretome of MSCs also contains various cytokines that contribute to scar reduction. Cytokines are small proteins that regulate immune responses and cell communication. Some of the cytokines involved in scar reduction include:

- Interleukin-10 (IL-10): IL-10 is an anti-inflammatory cytokine that suppresses the production of pro-inflammatory cytokines. By reducing inflammation, IL-10 promotes a favorable environment for scar reduction and tissue regeneration (Tsuji et al., 2022).

- Interleukin-6 (IL-6): IL-6 is a pleiotropic cytokine that has both pro-inflammatory and anti-inflammatory effects. In scar reduction, IL-6 plays a crucial role in modulating the inflammatory response, promoting tissue repair, and reducing scar formation (Tsuji et al., 2022).

- Interferon-gamma (IFN-γ): IFN-γ is a cytokine that regulates immune responses and has anti-fibrotic properties. In scar reduction, IFN-γ inhibits the production of collagen and other ECM components, leading to reduced scar formation and improved tissue healing (Merlo & Iacono, 2023).

Chemokines Involved in Scar Reduction

Chemokines are small signaling proteins that regulate the migration and activation of immune cells. They also play a role in scar reduction by modulating the inflammatory response and promoting tissue repair. Some of the chemokines involved in scar reduction include:

- C-X-C Motif Chemokine Ligand 12 (CXCL12): CXCL12 is a chemokine that attracts stem cells and immune cells to the site of injury. In scar reduction, CXCL12 promotes the recruitment of MSCs and other regenerative cells, leading to improved tissue healing and reduced scar formation (Merlo & Iacono, 2023).

- C-C Motif Chemokine Ligand 2 (CCL2): CCL2 is a chemokine that attracts monocytes and macrophages to the site of injury. In scar reduction, CCL2 plays a crucial role in modulating the inflammatory response and promoting tissue repair, leading to reduced scar formation (Merlo & Iacono, 2023).

- C-C Motif Chemokine Ligand 5 (CCL5): CCL5 is a chemokine that regulates the migration and activation of immune cells. In scar reduction, CCL5 promotes the recruitment of immune cells that are involved in tissue remodeling and scar reduction (Merlo & Iacono, 2023).

In conclusion, scar reduction is a complex process that involves the modulation of various cellular and molecular mechanisms. The secretome of mesenchymal stem cells contains a diverse array of growth factors, cytokines, and chemokines that contribute to scar reduction by promoting tissue regeneration, modulating the inflammatory response, and regulating the synthesis and remodeling of the extracellular matrix. Further research in this field holds great promise for developing novel therapeutic strategies for scar reduction.

Further insights into Growth Factors and Cytokines involved in Scar Reduction

Scarring is a natural part of the healing process, but excessive scarring can lead to functional and aesthetic complications. Mesenchymal stem cells (MSCs) have shown great potential in scar reduction due to their ability to modulate the wound healing process. One of the key mechanisms through which MSCs exert their scar-reducing effects is the secretion of various cytokines. In this section, we will explore the cytokines involved in scar reduction and their role in promoting tissue regeneration (Chernoff, n.d.).

Transforming Growth Factor-beta (TGF-β)

TGF-β is a multifunctional cytokine that plays a crucial role in scar formation. While it is involved in the early stages of wound healing, excessive TGF-β signaling can lead to the accumulation of extracellular matrix components, resulting in fibrosis and scar formation. However, MSCs have been shown to modulate TGF-β signaling and promote scar reduction. MSCs can secrete soluble factors that inhibit TGF-β signaling, such as soluble TGF-β receptors and decorin, which can prevent excessive collagen deposition and promote tissue regeneration (Chernoff, n.d.).

Interleukin-10 (IL-10)

IL-10 is an anti-inflammatory cytokine that has been shown to play a crucial role in scar reduction. It can suppress the production of pro-inflammatory cytokines and promote the resolution of inflammation. MSCs have been found to secrete IL-10, which can inhibit the activation of fibroblasts and reduce collagen production, leading to scar reduction. Additionally, IL-10 can also promote the differentiation of MSCs into myofibroblasts, which are involved in wound contraction and scar formation (Chernoff, n.d.).

Hepatocyte Growth Factor (HGF)

HGF is a growth factor that is involved in various biological processes, including tissue repair and regeneration. It has been shown to have anti-fibrotic effects and can promote scar reduction. MSCs can secrete HGF, which can inhibit the activation of fibroblasts and reduce collagen production. HGF can also stimulate the migration and proliferation of epithelial cells, promoting re-epithelialization and wound closure (Preda, n.d.).

Vascular Endothelial Growth Factor (VEGF)

VEGF is a potent angiogenic factor that plays a crucial role in scar reduction. It can promote the formation of new blood vessels, which is essential for tissue regeneration. MSCs can secrete VEGF, which can stimulate angiogenesis and improve blood supply to the wound area. This enhanced blood flow can facilitate the delivery of oxygen and nutrients to the site of injury, promoting tissue repair and reducing scar formation (Preda, n.d.).

Platelet-Derived Growth Factor (PDGF)

PDGF is a growth factor that is involved in various cellular processes, including cell proliferation and migration. It has been shown to play a crucial role in scar reduction by promoting the recruitment and activation of fibroblasts, which are involved in the synthesis of extracellular matrix components. MSCs can secrete PDGF, which can stimulate the migration and proliferation of fibroblasts, leading to the remodeling of the wound bed and scar reduction (Khan et al., n.d.).

Insulin-like Growth Factor-1 (IGF-1)

IGF-1 is a growth factor that is involved in cell growth, proliferation, and differentiation. It has been shown to have anti-fibrotic effects and can promote scar reduction. MSCs can secrete IGF-1, which can stimulate the migration and proliferation of fibroblasts, leading to the remodeling of the wound bed and scar reduction. Additionally, IGF-1 can also promote the differentiation of MSCs into myofibroblasts, which are involved in wound contraction and scar formation (Khan et al., n.d.).

Connective Tissue Growth Factor (CTGF)

CTGF is a growth factor that is involved in various cellular processes, including cell adhesion, migration, and proliferation. It has been shown to play a crucial role in scar formation. However, MSCs can secrete factors that can inhibit CTGF expression, leading to scar reduction. MSCs can also secrete factors that can degrade CTGF, preventing excessive collagen deposition and promoting tissue regeneration (Khan et al., n.d.).

In conclusion, the secretome of MSCs contains a variety of cytokines that are involved in scar reduction. These cytokines can modulate the wound healing process, inhibit excessive collagen deposition, promote tissue regeneration, and improve the overall outcome of scar formation. Further research is needed to fully understand the mechanisms underlying the scar-reducing effects of MSCs and their secretome, but the potential for scar reduction using MSC-based therapies is promising (Chernoff, n.d.; Khan et al., n.d.; Preda, n.d.).

Further insights into Chemokines involved in Scar Reduction

Scarring is a natural part of the healing process, but excessive scarring can lead to functional and aesthetic complications. Mesenchymal stem cells (MSCs) have shown promising potential in scar reduction due to their ability to modulate the wound healing process. In addition to growth factors and cytokines, chemokines play a crucial role in scar reduction by regulating cell migration, proliferation, and tissue remodeling (Han et al., 2022).

CXCL12 (SDF-1)

Chemokine (C-X-C motif) ligand 12, also known as stromal cell-derived factor 1 (SDF-1), is a key chemokine involved in scar reduction. CXCL12 promotes the recruitment of MSCs to the site of injury, facilitating tissue repair and regeneration. It also plays a role in angiogenesis, which is essential for scar reduction. CXCL12 enhances the migration and homing of MSCs to the wound site, where they can exert their regenerative effects (Zhang et al., 2023).

CCL2 (MCP-1)

Chemokine (C-C motif) ligand 2, also known as monocyte chemoattractant protein 1 (MCP-1), is another chemokine involved in scar reduction. CCL2 attracts monocytes and macrophages to the wound site, promoting tissue remodeling and reducing scar formation. Macrophages play a crucial role in the resolution of inflammation and the removal of cellular debris, which are essential for scar reduction. CCL2 helps in recruiting these immune cells to facilitate the healing process (Rivera et al., 2020).

CXCL10 (IP-10)

Chemokine (C-X-C motif) ligand 10, also known as interferon gamma-induced protein 10 (IP-10), is involved in scar reduction by regulating the inflammatory response. CXCL10 attracts immune cells, such as T cells and natural killer cells, to the wound site. These immune cells play a role in tissue remodeling and scar reduction by modulating the inflammatory environment. CXCL10 also promotes angiogenesis, which is crucial for scar reduction and tissue regeneration (Han et al., 2022).

CCL5 (RANTES)

Chemokine (C-C motif) ligand 5, also known as regulated upon activation, normal T cell expressed and secreted (RANTES), is involved in scar reduction by regulating cell migration and tissue remodeling.

CCL5 attracts immune cells, such as T cells and eosinophils, to the wound site. These immune cells contribute to the resolution of inflammation and the remodeling of the extracellular matrix, leading to scar reduction. CCL5 also plays a role in angiogenesis, which is essential for tissue regeneration (Han et al., 2022).

CXCL8 (IL-8)

Chemokine (C-X-C motif) ligand 8, also known as interleukin-8 (IL-8), is involved in scar reduction by regulating the inflammatory response and promoting tissue remodeling. CXCL8 attracts neutrophils to the wound site, which are important for the initial inflammatory response. Neutrophils help in clearing pathogens and cellular debris, facilitating the healing process. CXCL8 also plays a role in angiogenesis, which is crucial for scar reduction and tissue regeneration (Qazi et al., 2022).

CXCL1 (GRO-α)

Chemokine (C-X-C motif) ligand 1, also known as growth-regulated oncogene alpha (GRO-α), is involved in scar reduction by regulating the inflammatory response and promoting tissue remodeling. CXCL1 attracts neutrophils and fibroblasts to the wound site. Neutrophils contribute to the initial inflammatory response, while fibroblasts are responsible for collagen deposition and tissue remodeling. CXCL1 also plays a role in angiogenesis, which is essential for scar reduction and tissue regeneration (Qazi et al., 2022).

CXCL5 (ENA-78)

Chemokine (C-X-C motif) ligand 5, also known as epithelial-derived neutrophil-activating peptide 78 (ENA-78), is involved in scar reduction by regulating the inflammatory response and promoting tissue remodeling. CXCL5 attracts neutrophils to the wound site, which are important for the initial inflammatory response. Neutrophils help in clearing pathogens and cellular debris, facilitating the healing process. CXCL5 also plays a role in angiogenesis, which is crucial for scar reduction and tissue regeneration (Qazi et al., 2022).

CXCL14 (BRAK)

Chemokine (C-X-C motif) ligand 14, also known as breast and kidney-expressed chemokine (BRAK), is involved in scar reduction by regulating the inflammatory response and promoting tissue re-modeling. CXCL14 attracts immune cells, such as monocytes and macrophages, to the wound site. These immune cells contribute to the resolution of inflammation and the remodeling of the extracellular matrix, leading to scar reduction. CXCL14 also plays a role in angio-genesis, which is essential for tissue regeneration (Han et al., 2022).

In conclusion, chemokines play a crucial role in scar reduction by regulating cell migration, inflammation, and tissue remodeling. CXCL12, CCL2, CXCL10, CCL5, CXCL8, CXCL1, CXCL5, and CXCL14 are some of the chemokines involved in the scar reduc-tion process. Understanding the role of these chemokines can provide valuable insights into the mechanisms underlying scar reduction and help in the development of novel therapeutic strategies for scar man-agement (Zhang et al., 2023; Rivera et al., 2020; Han et al., 2022; Qazi et al., 2022).

CHAPTER FOUR

CELL PROLIFERATION AND THE SECRETOME

Introduction to Cell Proliferation

C ell proliferation is a fundamental process in the body that plays a crucial role in tissue growth, development, and repair. It involves the division and multiplication of cells to generate new cells. In the context of regenerative medicine, cell proliferation is of great importance as it contributes to the regeneration and repair of damaged tissues. Mesenchymal stem cells (MSCs) have been found to possess remarkable abilities to promote cell proliferation, making

them a promising tool in regenerative therapies (Jiang et al., 2023).C
ell proliferation is a fundamental process in the body that

Growth Factors involved in Cell Proliferation

Growth factors are signaling molecules that regulate various cellular
processes, including cell proliferation. MSCs secrete a wide range of
growth factors that have been shown to stimulate cell proliferation
in different tissues. Some of the key growth factors involved in cell
proliferation include:

- Epidermal Growth Factor (EGF): EGF is a potent growth
  factor that stimulates the proliferation of various cell types,
  including epithelial cells, fibroblasts, and endothelial cells. It
  plays a crucial role in tissue regeneration and wound healing
  (Hormozi et al., 2023).

- Fibroblast Growth Factor (FGF): FGFs are a family of
  growth factors that promote cell proliferation and tissue repair. They have been shown to stimulate the proliferation of
  fibroblasts, endothelial cells, and other cell types involved in
  tissue regeneration (Hormozi et al., 2023).

- Platelet-Derived Growth Factor (PDGF): PDGF is a growth
  factor that is released from platelets during the clotting
  process. It plays a critical role in wound healing by stimulating the proliferation of fibroblasts, smooth muscle cells, and
  endothelial cells (Hormozi et al., 2023).

- Insulin-like Growth Factor (IGF): IGF is a growth factor
  that promotes cell proliferation and survival. It is involved in
  tissue growth and repair and has been shown to enhance the

proliferation of various cell types, including fibroblasts and endothelial cells (Chen et al., 2019).

Cytokines involved in Cell Proliferation

Cytokines are small proteins that regulate immune responses and cellular communication. They also play a role in cell proliferation and tissue regeneration. MSCs secrete a variety of cytokines that can stimulate cell proliferation. Some of the cytokines involved in cell proliferation include:

- Interleukin-6 (IL-6): IL-6 is a cytokine that promotes cell proliferation and survival. It has been shown to enhance the proliferation of various cell types, including fibroblasts and endothelial cells (Hade et al., 2021).

- Interleukin-17 (IL-17): IL-17 is a cytokine that regulates immune responses and cell proliferation. It has been shown to enhance the proliferation of fibroblasts and endothelial cells (Hade et al., 2021).

- Transforming Growth Factor-beta (TGF-β): TGF-β is a multifunctional cytokine that regulates cell proliferation, differentiation, and migration. It has been shown to promote the proliferation of various cell types involved in tissue repair, including fibroblasts and endothelial cells (Hade et al., 2021).

- Interferon-gamma (IFN-γ): IFN-γ is a cytokine that regulates immune responses and cell proliferation. It has been shown to enhance the proliferation of fibroblasts and endothelial cells (Hade et al., 2021).

Chemokines involved in Cell Proliferation

Chemokines are a type of cytokine that play a role in cell migration and proliferation. They are involved in various cellular processes, including tissue repair and regeneration. Some of the chemokines involved in cell proliferation include:

- Stromal cell-derived factor-1 (SDF-1): SDF-1 is a chemokine that promotes cell proliferation and migration. It plays a crucial role in tissue repair and regeneration by stimulating the proliferation of various cell types, including fibroblasts and endothelial cells (Hade et al., 2021).

- Monocyte Chemoattractant Protein-1 (MCP-1): MCP-1 is a chemokine that attracts monocytes and promotes cell proliferation. It has been shown to enhance the proliferation of fibroblasts and endothelial cells (Hade et al., 2021).

- Growth-Regulated Oncogene-alpha (GRO-α): GRO-α is a chemokine that stimulates cell proliferation and migration. It plays a role in tissue repair and regeneration by promoting the proliferation of fibroblasts and endothelial cells (Hade et al., 2021).

- Interleukin-8 (IL-8): IL-8 is a chemokine that stimulates cell proliferation and migration. It plays a role in tissue repair and regeneration by promoting the proliferation of endothelial cells and fibroblasts (Hade et al., 2021).

In conclusion, cell proliferation is a critical process in tissue regeneration, and MSCs have been found to possess the ability to promote cell proliferation through the secretion of various growth factors, cytokines, and chemokines. These molecules play a crucial role in stimulating the proliferation of different cell types involved in tissue repair and regeneration. Harnessing the regenerative potential of MSCs and their secretome holds great promise for the development of novel therapies for various diseases and injuries (Jiang et al., 2023; Hormozi et al., 2023; Hade et al., 2021; Chen et al., 2019).

Further insights into Growth Factors, Cytokines, and Chemokines involved in Cell Proliferation

Cell proliferation is a fundamental process in tissue regeneration and repair. It involves the growth and division of cells to replace damaged or lost cells. Mesenchymal stem cells (MSCs) have been shown to play a crucial role in promoting cell proliferation through the secretion of various cytokines. In this section, we will explore the cytokines involved in cell proliferation and their significance in the healing process.

Transforming Growth Factor-Beta (TGF-β)

Transforming Growth Factor-Beta (TGF-β) is a multifunctional cytokine that regulates cell growth, differentiation, and migration. It plays a crucial role in promoting cell proliferation by stimulating the synthesis of extracellular matrix components and activating signaling pathways involved in cell cycle progression. TGF-β also promotes the recruitment of immune cells to the site of injury, facilitating tissue repair (Kwaan & Lindholm, 2019).

Insulin-like Growth Factor (IGF)

Insulin-like Growth Factor (IGF) is a family of growth factors that includes IGF-1 and IGF-2. These growth factors are known to stimulate cell proliferation and enhance tissue regeneration. IGF-1 promotes the synthesis of DNA and proteins, leading to increased cell division. It also stimulates the production of collagen, an essential component of the extracellular matrix (Ozaki et al., 2007).

Epidermal Growth Factor (EGF)

Epidermal Growth Factor (EGF) is a potent mitogen that stimulates cell proliferation and migration. It plays a crucial role in wound healing and tissue regeneration. EGF promotes the growth and division of various cell types, including epithelial cells, fibroblasts, and endothelial cells. It also enhances the synthesis of collagen and other extracellular matrix components (Ozaki et al., 2007).

Platelet-Derived Growth Factor (PDGF)

Platelet-Derived Growth Factor (PDGF) is a potent cytokine involved in cell proliferation and angiogenesis. It is released by platelets and various other cell types, including MSCs. PDGF stimulates the proliferation of fibroblasts, smooth muscle cells, and endothelial cells, promoting tissue repair and regeneration. It also plays a crucial role in the formation of new blood vessels (Ozaki et al., 2007).

Fibroblast Growth Factor (FGF)

Fibroblast Growth Factor (FGF) is a family of growth factors that includes FGF-2 and FGF-7. These growth factors are known to stimulate cell proliferation and tissue regeneration. FGF-2 promotes the growth and division of various cell types, including fibroblasts, endothelial cells, and keratinocytes. It also enhances the synthesis of collagen and other extracellular matrix components (Wang et al., 2020).

Vascular Endothelial Growth Factor (VEGF)

Vascular Endothelial Growth Factor (VEGF) is a potent cytokine involved in angiogenesis and cell proliferation. It promotes the growth of new blood vessels, facilitating the delivery of oxygen and nutrients to the site of injury. VEGF also stimulates the proliferation of endothelial cells and enhances the migration of various cell types involved in tissue repair (Wang et al., 2020).

Interleukins (ILs)

Interleukins (ILs) are a group of cytokines that regulate immune responses and cell proliferation. Several ILs, including IL-6, IL-8, and IL-11, have been shown to promote cell proliferation and tissue regeneration. IL-6 stimulates the growth and division of various cell types, including fibroblasts and endothelial cells. IL-8 enhances the migration of cells involved in wound healing, while IL-11 promotes the synthesis of extracellular matrix components (Khan et al., n.d.).

Hepatocyte Growth Factor (HGF)

Hepatocyte Growth Factor (HGF) is a cytokine that promotes cell proliferation and tissue regeneration. It stimulates the growth and division of various cell types, including hepatocytes, fibroblasts, and endothelial cells. HGF also enhances the migration of cells involved in tissue repair and angiogenesis (Khan et al., n.d.).

Chemokines

Chemokines are a group of small cytokines that play a crucial role in cell proliferation and tissue regeneration. Chemokines, such as CXCL12 and CCL2, promote the recruitment of immune cells to the site of injury, facilitating tissue repair. They also stimulate the growth and division of various cell types involved in the healing process (Khan et al., n.d.).

In conclusion, the secretome of MSCs contains a diverse array of cytokines that promote cell proliferation and tissue regeneration. These cytokines, including TGF-β, IGF, EGF, PDGF, FGF, VEGF, ILs, HGF, and chemokines, play crucial roles in stimulating cell growth, enhancing extracellular matrix synthesis, and promoting angiogenesis. Understanding the role of these cytokines in cell proliferation can provide valuable insights into the regenerative potential of MSCs and their therapeutic applications in various diseases and injuries (Wang et al., 2020; Ozaki et al., 2007; Kwaan & Lindholm, 2019; Khan et al., n.d.).

CHAPTER FIVE

IMMUNO-MODULATION AND THE SECRETOME

Introduction to Immunomodulation

Immunomodulation is a crucial aspect of the regenerative properties of mesenchymal stem cells (MSCs). The ability of MSCs to modulate the immune system plays a significant role in their therapeutic potential. In this section, we will explore the various growth factors, cytokines, and chemokines involved in the immunomodulatory effects of the secretome of MSCs (Soetjahjo, 2022; Teixeira et al., 2013).

Growth Factors involved in Immunomodulation

The secretome of MSCs contains a diverse array of growth factors that contribute to their immunomodulatory effects. These growth factors have the ability to regulate the activity of immune cells and promote a balanced immune response. Some of the key growth factors involved in immunomodulation include:The secretome of MSCs contains a diverse array of growth factors that

- Transforming Growth Factor-beta (TGF-beta): TGF-beta is a potent immunosuppressive factor secreted by MSCs. It inhibits the activation and proliferation of T cells and promotes the generation of regulatory T cells (Tregs), which play a crucial role in maintaining immune tolerance (Lunyak et al., 2017).

- Hepatocyte Growth Factor (HGF): HGF is known for its anti-inflammatory properties and can suppress the activation of immune cells such as macrophages and dendritic cells. It also promotes tissue repair and regeneration (Lunyak et al., 2017).

- Indoleamine 2,3-dioxygenase (IDO): IDO is an enzyme secreted by MSCs that can inhibit T cell proliferation and promote the generation of Tregs. It also has immunosuppressive effects by depleting tryptophan, an essential amino acid for T cell proliferation (Lunyak et al., 2017).

- Prostaglandin E2 (PGE2): PGE2 is a lipid mediator that is involved in the regulation of immune responses. It can inhibit the activation of T cells and promote the expansion of Tregs. PGE2 also has anti-inflammatory effects (Lunyak et al., 2017).

Cytokines involved in Immunomodulation

Cytokines are small proteins secreted by MSCs that play a crucial role in modulating immune responses. The secretome of MSCs contains several cytokines that contribute to their immunomodulatory effects. Some of the key cytokines involved in immunomodulation include:

Cytokines are small proteins secreted by MSCs that play a crucial role in modulating immune responses. The secretome of MSCs contains several cytokines that contribute to their immunomodulatory effects.

- Interleukin-10 (IL-10): IL-10 is an anti-inflammatory cytokine that can suppress the production of pro-inflammatory cytokines by immune cells. It also promotes the generation of Tregs and inhibits the activation of macrophages (Esteban-Blanco et al., n.d.).

- Interleukin-6 (IL-6): IL-6 has both pro-inflammatory and anti-inflammatory properties depending on the context. In the context of immunomodulation, IL-6 can promote the generation of Tregs and inhibit the activation of immune cells (Esteban-Blanco et al., n.d.).

- Interleukin-1 receptor antagonist (IL-1ra): IL-1ra is a natural inhibitor of the pro-inflammatory cytokine interleukin-1 (IL-1). It can suppress the activation of immune cells and reduce inflammation (Esteban-Blanco et al., n.d.).

- Interferon-gamma (IFN-gamma): IFN-gamma is a cytokine that plays a crucial role in regulating immune responses. It can modulate the activity of immune cells and promote an

anti-inflammatory environment (Esteban-Blanco et al., n.d
.).

Chemokines involved in Immunomodulation

Chemokines are small signaling proteins that play a crucial role in immune cell trafficking and recruitment. The secretome of MSCs contains various chemokines that contribute to their immunomodulatory effects. Some of the key chemokines involved in immunomodulation include:

- CCL2 (also known as MCP-1): CCL2 is a chemokine that can recruit monocytes and macrophages to sites of inflammation. It also has immunomodulatory effects by inhibiting the activation of immune cells (Esteban-Blanco et al., n.d.).

- CXCL8 (also known as IL-8): CXCL8 is a chemokine that can recruit neutrophils to sites of inflammation. It also has immunomodulatory effects by inhibiting the activation of immune cells (Esteban-Blanco et al., n.d.).

- CXCL9: CXCL9 is a chemokine that can recruit T cells to sites of inflammation. It plays a crucial role in regulating immune responses and promoting an anti-inflammatory environment (Esteban-Blanco et al., n.d.).

- CXCL10: CXCL10 is a chemokine that can recruit T cells and natural killer (NK) cells to sites of inflammation. It has immunomodulatory effects by regulating the activity of immune cells (Esteban-Blanco et al., n.d.).

In conclusion, the secretome of MSCs contains a diverse array of growth factors, cytokines, and chemokines that contribute to their immunomodulatory effects. These factors play a crucial role in regulating immune responses and promoting a balanced immune environment. Understanding the immunomodulatory properties of MSCs and their secretome is essential for harnessing their therapeutic potential in various diseases and conditions (Soetjahjo, 2022; Teixeira et al., 2013; Lunyak et al., 2017; Esteban-Blanco et al., n.d.).

Advancing Beyond the Basics of Immunomodulation

Immunomodulation refers to the ability of mesenchymal stem cells (MSCs) to regulate and modulate the immune response. This property of MSCs has gained significant attention in the field of regenerative medicine due to its potential therapeutic applications in various immune-related disorders. The immunomodulatory effects of MSCs are mediated by a complex interplay of growth factors, cytokines, and chemokines present in their secretome. In this section, we will explore a bit deeper growth factors, cytokines and chemokines involved in immunomodulation and their role in regulating the immune system.

Further Insights into Growth Factors and Cytokines involved in Immunomodulation

Transforming Growth Factor-beta (TGF-β)

TGF-β is a key growth factor involved in immunomodulation by MSCs. It plays a crucial role in suppressing the immune response by inhibiting the proliferation and activation of various immune cells, including T cells, B cells, and natural killer (NK) cells. TGF-β also promotes the differentiation of regulatory T cells (Tregs), which are essential for maintaining immune tolerance and preventing autoimmune reactions (Qian et al., 2015).

Interleukin-10 (IL-10)

IL-10 is an anti-inflammatory cytokine secreted by MSCs that exerts potent immunomodulatory effects. It inhibits the production of pro-inflammatory cytokines such as tumor necrosis factor-alpha (TNF-α), interleukin-1 beta (IL-1β), and interferon-gamma (IFN-γ). IL-10 also promotes the differentiation of anti-inflammatory immune cells, such as M2 macrophages, which play a crucial role in tissue repair and regeneration (Cao et al., 2017).

Indoleamine 2,3-dioxygenase (IDO)

IDO is an enzyme secreted by MSCs that catalyzes the breakdown of the amino acid tryptophan. This metabolic pathway leads to the production of kynurenine, which has immunosuppressive properties. IDO-mediated tryptophan depletion inhibits the proliferation of T cells and promotes the generation of Tregs, thereby suppressing the immune response (Pinheiro et al., 2020).

Prostaglandin E2 (PGE2)

PGE2 is a lipid mediator produced by MSCs that exhibits potent immunomodulatory effects. It inhibits the activation and proliferation of T cells and NK cells, while promoting the expansion of Tregs. PGE2 also suppresses the production of pro-inflammatory cytokines and chemokines, thereby dampening the immune response (Pinheiro et al., 2020).

Hepatocyte Growth Factor (HGF)

HGF is a growth factor secreted by MSCs that plays a crucial role in immunomodulation. It inhibits the activation and proliferation of T cells and NK cells, while promoting the expansion of Tregs. HGF also enhances the migration and homing of MSCs to sites of inflammation, facilitating their immunomodulatory effects (Qian et al., 2015).

Interleukin-6 (IL-6)

IL-6 is a pleiotropic cytokine secreted by MSCs that exhibits both pro-inflammatory and anti-inflammatory properties depending on the context. It can promote the differentiation of pro-inflammatory immune cells, such as Th17 cells, but also suppresses the activation and proliferation of T cells. IL-6 plays a crucial role in maintaining immune homeostasis and regulating the balance between pro-inflammatory and anti-inflammatory responses (Cao et al., 2017).

Vascular Endothelial Growth Factor (VEGF)

VEGF is a growth factor secreted by MSCs that plays a vital role in angiogenesis, but it also exhibits immunomodulatory effects. VEGF promotes the recruitment and activation of immune cells involved in tissue repair and regeneration, such as macrophages and endothelial progenitor cells. It also enhances the production of anti-inflammatory cytokines, thereby modulating the immune response (Pinheiro et al., 2020).

Insulin-like Growth Factor-1 (IGF-1)

IGF-1 is a growth factor secreted by MSCs that regulates cell growth, proliferation, and differentiation. It also exhibits immunomodulatory effects by inhibiting the activation and proliferation of T cells and NK cells. IGF-1 promotes the differentiation of anti-inflammatory immune cells, such as M2 macrophages, and enhances tissue repair and regeneration (Pinheiro et al., 2020).

These are just a few examples of the growth factors involved in immunomodulation by MSCs. The secretome of MSCs contains a diverse array of bioactive molecules that collectively contribute to their immunomodulatory properties. Further research is needed to fully understand the mechanisms underlying the immunomodulatory effects of MSCs and to harness their therapeutic potential in immune-related disorders (Qian et al., 2015; Cao et al., 2017; Pinheiro et al., 2020).

Further insights into Chemokines involved in Immunomodulation

Chemokines are a group of small proteins that play a crucial role in immunomodulation. They are involved in the recruitment and activation of immune cells, helping to regulate the immune response. Mesenchymal stem cells (MSCs) have been found to secrete various chemokines that contribute to their immunomodulatory effects. In this section, we will explore the chemokines involved in immunomodulation by MSCs and their potential therapeutic applications (Mishra & Panwar, 2023).

CCL2 (C-C Motif Chemokine Ligand 2)

CCL2, also known as monocyte chemoattractant protein-1 (MCP-1), is a chemokine that plays a crucial role in the recruitment of monocytes and macrophages. MSCs have been shown to secrete CCL2, which can attract monocytes to the site of inflammation. Once recruited, monocytes can differentiate into anti-inflammatory M2 macrophages, promoting tissue repair and reducing inflammation (Patrikoski et al., 2014).

CCL5 (C-C Motif Chemokine Ligand 5)

CCL5, also known as regulated upon activation, normal T cell expressed and secreted (RANTES), is a chemokine involved in the recruitment and activation of various immune cells, including T cells, monocytes, and eosinophils. MSCs have been found to secrete CCL5, which can attract immune cells to the site of injury or inflammation. This recruitment of immune cells can help modulate the immune response and promote tissue regeneration (Patrikoski et al., 2014).

CXCL8 (C-X-C Motif Chemokine Ligand 8)

CXCL8, also known as interleukin-8 (IL-8), is a chemokine that plays a crucial role in the recruitment and activation of neutrophils. MSCs have been shown to secrete CXCL8, which can attract neutrophils to the site of inflammation. Neutrophils are important immune cells involved in the early stages of the immune response. By recruiting neutrophils, MSCs can help regulate the immune response and promote tissue healing (Patrikoski et al., 2014).

CXCL9 (C-X-C Motif Chemokine Ligand 9) and CXCL10 (C-X-C Motif Chemokine Ligand 10)

CXCL9 and CXCL10 are chemokines involved in the recruitment and activation of T cells. MSCs have been found to secrete CXCL9 and CXCL10, which can attract T cells to the site of inflammation. T cells play a crucial role in the immune response, and their recruitment by MSCs can help modulate the immune system and promote tissue repair (Tarte et al., 2005).

CX3CL1 (C-X3-C Motif Chemokine Ligand 1)

CX3CL1, also known as fractalkine, is a chemokine that plays a role in the recruitment and activation of immune cells, including monocytes, T cells, and natural killer (NK) cells. MSCs have been shown to secrete CX3CL1, which can attract immune cells to the site of injury or inflammation. This recruitment of immune cells can help regulate the immune response and promote tissue regeneration (Tarte et al., 2005).

Other Chemokines

In addition to the chemokines mentioned above, MSCs have been found to secrete various other chemokines, including CCL20, CXCL1, CXCL2, and CXCL12. These chemokines play important roles in immune cell recruitment and activation, contributing to the immunomodulatory effects of MSCs (Mishra & Panwar, 2023).

The secretion of these chemokines by MSCs helps to regulate the immune response and promote tissue repair. By attracting immune cells to the site of injury or inflammation, MSCs can modulate the immune system and create a favorable environment for tissue regeneration. The immunomodulatory properties of MSCs and their secretome make them promising candidates for various therapeutic applications, including the treatment of autoimmune diseases, graft-versus-host disease, and tissue injuries (Mishra & Panwar, 2023; Patrikoski et al., 2014; Tarte et al., 2005).

Further research is needed to fully understand the mechanisms by which MSCs and their secretome exert their immunomodulatory effects. However, the identification of specific chemokines involved in immunomodulation provides valuable insights into the potential therapeutic applications of MSCs in immune-related disorders. Harnessing the power of MSCs and their secretome may pave the way for innovative and effective treatments in the field of regenerative medicine.

CHAPTER SIX

ANTI-INFLAMMATORY EFFECTS OF THE SECRETOME

Introduction to Anti-inflammatory Effects

Inflammation is a natural response of the body to injury or infection. It is characterized by redness, swelling, heat, and pain. While acute inflammation is a necessary part of the healing process, chronic inflammation can lead to tissue damage and contribute to the development of various diseases. Inflammation is a natural response of the body to injury or infection.

The remarkable anti-inflammatory properties of Mesenchymal stem cells (MSCs) have garnered considerable interest within the realm of regenerative medicine.

These effects are primarily mediated by the secretome of MSCs, which consists of a complex mixture of bioactive molecules including growth factors, cytokines, and chemokines (Jafri et al., 2022; Adas et al., 2021).

Growth Factors involved in Anti-inflammatory Effects

Several growth factors present in the secretome of MSCs contribute to their anti-inflammatory effects. One of the key growth factors is transforming growth factor-beta (TGF-β), which plays a crucial role in regulating immune responses and promoting tissue repair. TGF-β has been shown to inhibit the production of pro-inflammatory cytokines and promote the differentiation of immune cells into anti-inflammatory phenotypes (Jafri et al., 2022).

Another important growth factor is hepatocyte growth factor (HGF), which has potent anti-inflammatory properties. HGF can suppress the activation of immune cells and inhibit the production of pro-inflammatory molecules. It also promotes tissue regeneration and repair (Ziegler et al., 2019).

Vascular endothelial growth factor (VEGF) is another growth factor present in the secretome of MSCs that exhibits anti-inflammatory effects. VEGF can reduce inflammation by promoting the growth of new blood vessels, improving tissue perfusion, and facilitating the removal of inflammatory mediators (Jafri et al., 2022).

Cytokines involved in Anti-inflammatory Effects

The secretome of MSCs contains various cytokines that contribute to their anti-inflammatory effects. Interleukin-10 (IL-10) is a key cytokine with potent anti-inflammatory properties. IL-10 can suppress the production of pro-inflammatory cytokines and promote the differentiation of immune cells into anti-inflammatory phenotypes (Adas et al., 2021).

Another important cytokine is interleukin-1 receptor antagonist (IL-1Ra), which acts as a competitive inhibitor of pro-inflammatory cytokines such as interleukin-1 (IL-1). IL-1Ra can effectively dampen the inflammatory response and reduce tissue damage (Ziegler et al., 2019).

Interleukin-6 (IL-6) is a multifunctional cytokine that exhibits both pro-inflammatory and anti-inflammatory properties. In the context of MSC secretome, IL-6 has been shown to have anti-inflammatory effects by inhibiting the production of pro-inflammatory cytokines and promoting the differentiation of immune cells into anti-inflammatory phenotypes (Jafri et al., 2022).

Chemokines involved in Anti-inflammatory Effects

Chemokines present in the secretome of MSCs also play a role in their anti-inflammatory effects. One of the key chemokines is stromal cell-derived factor-1 (SDF-1), also known as CXCL12. SDF-1 can recruit immune cells with anti-inflammatory properties and promote tissue repair (Jafri et al., 2022).

Another important chemokine is monocyte chemoattractant protein-1 (MCP-1), also known as CCL2. MCP-1 can attract immune cells with anti-inflammatory phenotypes and promote the resolution of inflammation (Ziegler et al., 2019).

Overall, the secretome of MSCs exerts potent anti-inflammatory effects through the action of various growth factors, cytokines, and chemokines. These molecules work in a coordinated manner to modulate the immune response, suppress inflammation, and promote tissue repair. Harnessing the therapeutic potential of the secretome holds great promise for the development of novel anti-inflammatory therapies in the field of regenerative medicine (Jafri et al., 2022; Ziegler et al., 2019; Adas et al., 2021).

Foundations of Anti-Inflammatory Mechanisms of MSC

In addition to their regenerative properties, mesenchymal stem cells (MSCs) have been found to possess potent anti-inflammatory effects. These effects are primarily mediated by the secretion of various growth factors, cytokines, and chemokines, collectively known as the secretome. The secretome of MSCs plays a crucial role in modulating the immune response and reducing inflammation in various disease conditions (Chaudhary et al., 2022).

Further Insights into Growth Factors and Cytokines involved in Anti-Inflammatory Mechanisms

Transforming Growth Factor-beta (TGF-β)

One of the key growth factors involved in the anti-inflammatory effects of the secretome is Transforming Growth Factor-beta (TGF-β). TGF-β is known to have immunosuppressive properties and can inhibit the activation and proliferation of immune cells such as T cells and B cells. It also promotes the differentiation of regulatory T cells (Tregs), which play a crucial role in suppressing excessive immune responses and maintaining immune homeostasis (Raj et al., 2021).

Interleukin-10 (IL-10)

Interleukin-10 (IL-10) is a cytokine that is secreted by MSCs and has potent anti-inflammatory effects. IL-10 acts by inhibiting the production of pro-inflammatory cytokines such as tumor necrosis factor-alpha (TNF-α), interleukin-1 beta (IL-1β), and interleukin-6 (IL-6). It also suppresses the activation of immune cells and promotes the differentiation of anti-inflammatory immune cells, such as M2 macrophages (Shandil et al., 2022).

Hepatocyte Growth Factor (HGF)

Hepatocyte Growth Factor (HGF) is another growth factor secreted by MSCs that exhibits anti-inflammatory properties. HGF has been shown to inhibit the production of pro-inflammatory cytokines and chemokines, such as TNF-α, IL-1β, and monocyte chemoattractant protein-1 (MCP-1). It also promotes tissue repair and regeneration by stimulating the migration and proliferation of various cell types, including endothelial cells and fibroblasts (Jiao et al., 2021).

Indoleamine 2,3-Dioxygenase (IDO)

Indoleamine 2,3-Dioxygenase (IDO) is an enzyme that is upregulated in MSCs upon exposure to inflammatory stimuli. IDO plays a crucial r le in the immunomodulatory effects of MSCs by catalyzing the breakdown of tryptophan, an essential amino acid required for T cell proliferation. By depleting tryptophan, IDO inhibits T cell activation and promotes immune tolerance (Shandil et al., 2022).

Prostaglandin E2 (PGE2)

Prostaglandin E2 (PGE2) is a lipid mediator that is secreted by MSCs and exhibits potent anti-inflammatory effects. PGE2 acts by inhibiting the production of pro-inflammatory cytokines, such as TNF-α and IL-1β, and promoting the production of anti-inflammatory cytokines, such as IL-10. It also suppresses the activation of immune cells and promotes the differentiation of regulatory immune cells, such as Tregs (Raj et al., 2021).

Interleukin-10 (IL-10)

Interleukin-10 (IL-10) is a key cytokine involved in the anti-inflammatory effects of MSCs. It is known for its ability to suppress the production of pro-inflammatory cytokines and promote the production of anti-inflammatory cytokines. IL-10 acts by inhibiting the activation and function of immune cells, such as macrophages and T cells, that are responsible for initiating and sustaining the inflammatory response. By reducing the production of pro-inflammatory cytokines, IL-10 helps to dampen the inflammatory cascade and promote tissue healing (Karimi et al., 2021).

Transforming Growth Factor-beta (TGF-β)

Transforming Growth Factor-beta (TGF-β) is another important cytokine involved in the anti-inflammatory effects of MSCs. TGF-β has diverse functions in the immune system, including the regulation of immune cell differentiation and the suppression of immune responses. In the context of inflammation, TGF-β acts to inhibit the activation and proliferation of immune cells, thereby reducing the production of pro-inflammatory cytokines. Additionally, TGF-β promotes the differentiation of regulatory T cells, which play a crucial role in suppressing excessive immune responses and maintaining immune homeostasis (Yi et al., 2020).

Interleukin-1 Receptor Antagonist (IL-1Ra)

Interleukin-1 Receptor Antagonist (IL-1Ra) is a cytokine that acts as a natural inhibitor of interleukin-1 (IL-1), a pro-inflammatory cytokine involved in the initiation and amplification of the inflammatory response. IL-1Ra competes with IL-1 for binding to its receptor, thereby preventing IL-1 signaling and reducing the inflammatory response. MSCs have been found to secrete IL-1Ra, which can effectively counteract the pro-inflammatory effects of IL-1 and contribute to the overall anti-inflammatory environment (Abdul-Hamid et al., 2023).

Interleukin-4 (IL-4)

Interleukin-4 (IL-4) is a cytokine that plays a crucial role in regulating the immune response and promoting anti-inflammatory effects. IL-4 is known for its ability to suppress the production of pro-inflammatory cytokines and promote the differentiation of immune cells towards an anti-inflammatory phenotype. MSCs have been shown to secrete IL-4, which can modulate the immune response by inhibiting the activation of immune cells and promoting the production of anti-inflammatory cytokines (Mani et al., 2023).

Interleukin-6 (IL-6)

Interleukin-6 (IL-6) is a multifunctional cytokine that can have both pro-inflammatory and anti-inflammatory effects depending on the context. In the case of MSCs, IL-6 has been found to exert anti-inflammatory effects by inhibiting the production of pro-inflammatory cytokines and promoting the differentiation of immune cells towards an anti-inflammatory phenotype. Additionally, IL-6 can stimulate the production of IL-10, further enhancing the anti-inflammatory properties of MSCs (Karimi et al., 2021).

Interleukin-13 (IL-13)

Interleukin-13 (IL-13) is a cytokine that shares similarities with IL-4 in terms of its anti-inflammatory properties. IL-13 can suppress the production of pro-inflammatory cytokines and promote the production of anti-inflammatory cytokines. MSCs have been shown to secrete IL-13, which can modulate the immune response and contribute to the overall anti-inflammatory effects of MSCs (Yi et al., 2020).

Interferon-gamma (IFN-γ)

Interferon-gamma (IFN-γ) is a cytokine that plays a complex role in inflammation and immune regulation. While IFN-γ is generally considered a pro-inflammatory cytokine, it can also have anti-inflammatory effects under certain conditions. MSCs have been found to secrete IFN-γ, which can modulate the immune response by inhibiting the activation of immune cells and promoting the production of anti-inflammatory cytokines (Abdul-Hamid et al., 2023).

Other Factors

In addition to the above-mentioned growth factors, MSCs secrete various other factors that contribute to their anti-inflammatory effects. These include interleukin-1 receptor antagonist (IL-1Ra), which competitively inhibits the binding of IL-1 to its receptor and prevents pro-inflammatory signaling. MSCs also secrete soluble TNF receptors (sTNFRs), which neutralize the activity of TNF-α, a key pro-inflammatory cytokine (Jiao et al., 2021).

In addition to the cytokines mentioned above, MSCs secrete various other cytokines that contribute to their anti-inflammatory effects. These include but are not limited to Interleukin-11 (IL-11), Interleukin-1 Receptor Type 2 (IL-1R2), Interleukin-1 Receptor Type 4 (IL-1R4), Interleukin-1 Receptor Type 5 (IL-1R5), Interleukin-1 Receptor Type 6 (IL-1R6), Interleukin-1 Receptor Type 7 (IL-1R7), Interleukin-1 Receptor Type 8 (IL-1R8), Interleukin-1 Receptor Type 9 (IL-1R9), Interleukin-1 Receptor Type 10 (IL-1R10), and Interleukin-1 Receptor Type 11 (IL-1R11). These cytokines collectively contribute to the modulation of the immune response and the promotion of an anti-inflammatory environment (Mani et al., 2023; Karimi et al., 2021; Yi et al., 2020).

The cytokines secreted by MSCs play a crucial role in mediating their anti-inflammatory effects. By modulating the immune response and reducing inflammation, these cytokines contribute to the regenerative properties of MSCs and their potential therapeutic applications in various inflammatory conditions. Further research is still needed to fully understand the mechanisms underlying the anti-inflammatory effects of MSCs and their secretome, but the current evidence suggests that cytokines play a significant role in this process (Abdul-Hamid et al., 2023; Mani et al., 2023; Karimi et al., 2021; Yi et al., 2020).

Furthermore, MSCs secrete various chemokines, such as stromal-derived factor-1 (SDF-1) and monocyte chemoattractant protein-3 (MCP-3), which play a role in recruiting immune cells to the site of inflammation and modulating their function. These chemokines can attract immune cells with anti-inflammatory properties, such as M2 macrophages and Tregs, thereby promoting an anti-inflammatory environment (Chaudhary et al., 2022).

In conclusion, the secretome of MSCs contains a diverse array of growth factors, cytokines, and chemokines that contribute to their anti-inflammatory effects. These factors act synergistically to suppress excessive immune responses, inhibit the production of pro-inflammatory molecules, and promote the differentiation of anti-inflammatory immune cells. Understanding the role of these factors in the anti-inflammatory effects of MSCs is crucial for harnessing their therapeutic potential in various inflammatory diseases (Raj et al., 2021; Jiao et al., 2021; Shandil et al., 2022; Chaudhary et al., 2022).

Further Insights into Chemokines involved in Anti-inflammatory Effects

Chemokines are a group of small proteins that play a crucial role in the immune system by regulating the migration and activation of immune cells. In the context of the secretome of Mesenchymal Stem Cells (MSCs), certain chemokines have been found to have anti-inflammatory effects. These chemokines can modulate the immune response and help reduce inflammation in various tissues and organs (Darwish et al., 2013).

CXCL8 (Interleukin-8)

CXCL8, also known as Interleukin-8, is a chemokine that is secreted by MSCs and has potent anti-inflammatory properties. It acts as a chemoattractant for neutrophils, monocytes, and T cells, which are important immune cells involved in the inflammatory response. By recruiting these immune cells to the site of inflammation, CXCL8 helps to regulate the immune response and promote tissue healing (Shabgah et al., 2019).

Studies have shown that MSCs can secrete CXCL8 in response to inflammatory stimuli, such as pro-inflammatory cytokines. CXCL8 acts by binding to its receptors on immune cells, leading to the inhibition of their activation and the suppression of pro-inflammatory cytokine production. This anti-inflammatory effect of CXCL8 helps to dampen the immune response and reduce tissue damage caused by inflammation (Shabgah et al., 2019).

CCL2 (Monocyte Chemoattractant Protein-1)

CCL2, also known as Monocyte Chemoattractant Protein-1, is another chemokine secreted by MSCs that exhibits anti-inflammatory effects. CCL2 plays a crucial role in the recruitment of monocytes, a type of immune cell, to sites of inflammation. By attracting monocytes to the inflamed tissue, CCL2 helps to regulate the immune response and promote tissue repair (Panahipour et al., 2022).

MSCs can secrete CCL2 in response to inflammatory signals, such as pro-inflammatory cytokines and chemokines. CCL2 acts by binding to its receptors on monocytes, leading to their migration to the site of inflammation. Once at the site, monocytes can differentiate into macrophages, which are important immune cells involved in tissue repair and inflammation resolution. The presence of CCL2 helps to modulate the immune response and promote the resolution of inflammation (Panahipour et al., 2022).

CXCL10 (Interferon-gamma Induced Protein 10)

CXCL10, also known as Interferon-gamma Induced Protein 10, is a chemokine that is secreted by MSCs and has anti-inflammatory properties. CXCL10 is primarily involved in the recruitment and activation of T cells, which are important immune cells involved in the immune response. By regulating T cell migration and activation, CXCL10 helps to modulate the immune response and reduce inflammation (Germano et al., 2010).

MSCs can secrete CXCL10 in response to inflammatory signals, such as interferon-gamma. CXCL10 acts by binding to its receptors on T cells, leading to the inhibition of their migration and activation. This anti-inflammatory effect of CXCL10 helps to suppress the immune response and reduce tissue inflammation (Germano et al., 2010).

CCL5 (Regulated upon Activation, Normal T-cell Expressed, and Secreted)

CCL5, also known as Regulated upon Activation, Normal T-cell Expressed, and Secreted (RANTES), is a chemokine secreted by MSCs that exhibits anti-inflammatory effects. CCL5 is involved in the recruitment and activation of various immune cells, including T cells, monocytes, and eosinophils. By regulating the migration and activation of these immune cells, CCL5 helps to modulate the immune response and reduce inflammation (Darwish et al., 2013).

MSCs can secrete CCL5 in response to inflammatory stimuli, such as pro-inflammatory cytokines and chemokines. CCL5 acts by binding to its receptors on immune cells, leading to the inhibition of their migration and activation. This anti-inflammatory effect of CCL5 helps to dampen the immune response and reduce tissue inflammation (Darwish et al., 2013).

CXCL12 (Stromal Cell-Derived Factor-1)

CXCL12, also known as Stromal Cell-Derived Factor-1 (SDF-1), is a chemokine secreted by MSCs that has anti-inflammatory properties. CXCL12 is involved in the recruitment and retention of immune cells, such as T cells and monocytes, to sites of inflammation. By regulating the migration and retention of these immune cells, CXCL12 helps to modulate the immune response and reduce inflammation (Germano et al., 2010).

MSCs can secrete CXCL12 in response to inflammatory signals, such as pro-inflammatory cytokines and chemokines. CXCL12 acts by binding to its receptors on immune cells, leading to the inhibition of their migration and retention at the site of inflammation. This anti-inflammatory effect of CXCL12 helps to suppress the immune response and reduce tissue inflammation (Germano et al., 2010).

Other Chemokines

In addition to the chemokines mentioned above, MSCs can secrete various other chemokines that have anti-inflammatory effects. These include CCL20, CXCL9, CXCL11, and CX3CL1. Each of these chemokines plays a role in modulating the immune response and reducing inflammation in different tissues and organs (Darwish et al., 2013).

The secretion of these chemokines by MSCs is tightly regulated and can be influenced by various factors, such as the presence of inflammatory signals and the microenvironment of the tissue. By secreting these anti-inflammatory chemokines, MSCs contribute to the resolution of inflammation and the promotion of tissue healing (Darwish et al., 2013).

In conclusion, the secretome of MSCs contains several chemokines that have anti-inflammatory effects. These chemokines help to modulate the immune response and reduce inflammation by regulating the migration and activation of immune cells. By understanding the role of these chemokines, researchers can further explore the therapeutic potential of MSCs and their secretome in the treatment of inflammatory diseases and tissue injuries (Shabgah et al., 2019; Panahipour et al., 2022; Germano et al., 2010; Darwish et al., 2013).

CHAPTER SEVEN

ANTI-MICROBIAL EFFECTS OF THE SECRETOME

Growth Factors involved in Anti-microbial Effects

The secretome of Mesenchymal Stem Cells (MSCs) is a complex mixture of various bioactive molecules, including growth factors, cytokines, and chemokines. These components play a crucial role in the therapeutic effects of MSCs, including their anti-microbial properties. In this section, we will explore the growth factors involved in the anti-microbial effects of the secretome.

Defensins

Defensins are small cationic peptides that exhibit potent antimicrobial activity. They are one of the key components of the secretome responsible for the anti-microbial effects of MSCs. Defensins can directly kill a wide range of microorganisms, including bacteria, fungi, and viruses. They exert their antimicrobial activity by disrupting the microbial cell membrane and inhibiting microbial growth (Tecle, Tripathi, & Hartshorn, 2010).

Cathelicidins

Cathelicidins are another group of antimicrobial peptides found in the secretome of MSCs. These peptides possess broad-spectrum antimicrobial activity against various pathogens, including bacteria, fungi, and viruses. Cathelicidins exert their antimicrobial effects by disrupting the microbial cell membrane, promoting the clearance of pathogens, and modulating the immune response (Chung, 2013).

LL-37

LL-37 is a specific cathelicidin peptide that plays a crucial role in the anti-microbial effects of the secretome. It exhibits potent antimicrobial activity against a wide range of microorganisms, including drug-resistant bacteria. LL-37 not only directly kills pathogens but also enhances the immune response against infections. It can recruit immune cells to the site of infection and promote the clearance of pathogens (Krasnodembskaya et al., 2010).

Hepcidin

Hepcidin is a peptide hormone that regulates iron metabolism and exhibits antimicrobial activity. It is involved in the defense against bacterial and fungal infections. Hepcidin exerts its antimicrobial effects by sequestering iron, an essential nutrient for microbial growth, and limiting its availability to pathogens. By depriving pathogens of iron, hepcidin inhibits their growth and replication (Silva-Carvalho et al., 2021).

Granulocyte-Macrophage Colony-Stimulating Factor (GM-CSF)

GM-CSF is a growth factor that stimulates the production and activation of immune cells, including neutrophils, macrophages, and dendritic cells. It plays a crucial role in the immune response against microbial infections. GM-CSF enhances the phagocytic activity of immune cells, promotes the production of antimicrobial molecules, and facilitates the clearance of pathogens (Ghosh & Weinberg, 2021).

Interferons (IFNs)

Interferons are a group of cytokines that have potent antiviral and immunomodulatory effects. They are involved in the defense against viral infections and play a crucial role in the anti-microbial effects of the secretome. Interferons inhibit viral replication, enhance the immune response against viruses, and modulate the activity of immune cells (Silva-Carvalho et al., 2021).

Tumor Necrosis Factor-Alpha (TNF-α)

TNF-α is a pro-inflammatory cytokine that plays a dual role in the immune response against microbial infections. It can promote inflammation and recruit immune cells to the site of infection. TNF-α also exhibits direct antimicrobial activity by inducing the production of reactive oxygen species (ROS) and antimicrobial peptides. However, excessive TNF-α production can lead to tissue damage and inflammation (Silva-Carvalho et al., 2021).

Transforming Growth Factor-Beta (TGF-β)

TGF-β is a multifunctional cytokine that regulates various cellular processes, including immune responses and tissue repair. It plays a crucial role in the anti-microbial effects of the secretome by modulating the activity of immune cells. TGF-β can suppress the immune response against pathogens, prevent excessive inflammation, and promote tissue healing (Silva-Carvalho et al., 2021).

Vascular Endothelial Growth Factor (VEGF)

VEGF is a growth factor that promotes the growth and repair of blood vessels. It plays a crucial role in angiogenesis, which is the formation of new blood vessels. While angiogenesis is primarily associated with tissue repair and regeneration, it also plays a role in the immune response against microbial infections. By promoting the formation of new blood vessels, VEGF facilitates the delivery of immune cells and antimicrobial molecules to the site of infection (Silva-Carvalho et al., 2021).

Hepatocyte Growth Factor (HGF)

Hepatocyte growth factor is a growth factor that is involved in tissue repair and regeneration. It is produced by various cell types, including MSCs. HGF has been shown to enhance the anti-microbial effects of MSCs by promoting the production of antimicrobial peptides and modulating the immune response to microbial infections (Kyurkchiev et al., 2014).

Granulocyte-Macrophage Colony-Stimulating Factor (GM-CSF)

Granulocyte-macrophage colony-stimulating factor is a growth factor that is involved in the production and activation of immune cells, such as neutrophils and macrophages. It is produced by various cell types, including immune cells and MSCs themselves. GM-CSF has been shown to enhance the anti-microbial effects of MSCs by promoting the recruitment and activation of immune cells at the site of infection (Naserian et al., 2020).

These are just a few examples of the growth factors involved in the anti-microbial effects of the secretome. The secretome of MSCs is a complex mixture of various bioactive molecules, and further research is needed to fully understand the mechanisms underlying its anti-microbial properties. Nonetheless, the presence of these growth factors in the secretome highlights the potential of MSCs as a therapeutic tool for combating microbial infections.

Cytokines involved in Anti-microbial Effects

The secretome of mesenchymal stem cells (MSCs) is a complex mixture of various bioactive molecules, including growth factors, cytokines, and chemokines. These components play a crucial role in the therapeutic effects of MSCs, including their anti-microbial properties (Madrigal, Rao, & Riordan, 2014). In this section, we will explore the cytokines involved in the anti-microbial effects of the secretome.

Interferon-gamma (IFN-γ)

Interferon-gamma is a cytokine that plays a vital role in the immune response against microbial infections. It is produced by various immune cells, including T cells and natural killer cells. IFN-γ has been shown to enhance the anti-microbial activity of MSCs by promoting the production of antimicrobial peptides and enhancing phagocytosis by immune cells (Weiss & Dahlke, 2019).

Tumor Necrosis Factor-alpha (TNF-α)

Tumor necrosis factor-alpha is a pro-inflammatory cytokine that is involved in the immune response against microbial pathogens. It is produced by immune cells such as macrophages and T cells. TNF-α has been shown to enhance the anti-microbial effects of MSCs by promoting the production of antimicrobial peptides and increasing the recruitment and activation of immune cells at the site of infection (Prockop & Oh, 2012).

Interleukin-1beta (IL-1β)

Interleukin-1beta is a pro-inflammatory cytokine that is produced by immune cells in response to microbial infections. It plays a crucial role in the activation of immune cells and the initiation of the inflammatory response. IL-1β has been shown to enhance the anti-microbial effects of MSCs by promoting the production of antimicrobial peptides and increasing the recruitment and activation of immune cells (Shigdar et al., 2014).

Interleukin-6 (IL-6)

Interleukin-6 is a cytokine that is involved in the regulation of the immune response and inflammation. It is produced by various immune cells, including macrophages and T cells. IL-6 has been shown to enhance the anti-microbial effects of MSCs by promoting the production of antimicrobial peptides and increasing the recruitment and activation of immune cells (Ranganath, Levy, Inamdar, & Karp, 2012).

Interleukin-8 (IL-8)

Interleukin-8 is a chemokine that is involved in the recruitment and activation of immune cells at the site of infection. It is produced by various cell types, including immune cells and MSCs themselves. IL-8 has been shown to enhance the anti-microbial effects of MSCs by promoting the recruitment and activation of immune cells, such as neutrophils, which are important in the defense against microbial pathogens (Brandau et al., 2014).

Transforming Growth Factor-beta (TGF-β)

Transforming growth factor-beta is a cytokine that plays a crucial role in tissue repair and regeneration. It is produced by various cell types, including immune cells and MSCs themselves. TGF-β has been shown to enhance the anti-microbial effects of MSCs by promoting the production of antimicrobial peptides and modulating the immune response to microbial infections (Konala, Mamidi, Bhonde, Das, Pochampally, & Pal, 2016).

These cytokines, along with other components of the secretome, work together to enhance the anti-microbial effects of MSCs. By promoting the production of antimicrobial peptides, modulating the immune response, and recruiting and activating immune cells, MSCs contribute to the defense against microbial pathogens. Understanding the role of these cytokines in the anti-microbial effects of MSCs is crucial for harnessing the therapeutic potential of MSCs in the treatment of infectious diseases.

Chemokines involved in Anti-microbial Effects

Chemokines are small proteins that play a crucial role in the immune response by attracting immune cells to the site of infection or inflammation. In addition to their role in immune cell recruitment, chemokines also have antimicrobial properties. The secretome of Mesenchymal Stem Cells (MSCs) contains various chemokines that contribute to their anti-microbial effects. In this section, we will explore the chemokines involved in the anti-microbial effects of the secretome and their potential therapeutic applications.

Regulated upon Activation, Normal T-cell Expressed, and Secreted (RANTES)

Regulated upon activation, normal T-cell expressed, and secreted, also known as CCL5, is a chemokine that is involved in the recruitment and activation of immune cells. It is produced by various cell types, including immune cells and MSCs themselves. RANTES has been shown to enhance the anti-microbial effects of MSCs by promoting the recruitment and activation of immune cells at the site of infection (Asami et al., 2013).

Stromal-Derived Factor-1 (SDF-1)

Stromal-derived factor-1, also known as CXCL12, is a chemokine that is involved in the recruitment and migration of immune cells. It is produced by various cell types, including MSCs. SDF-1 has been shown to enhance the anti-microbial effects of MSCs by promoting the recruitment and activation of immune cells, such as neutrophils and macrophages, at the site of infection (Lee et al., 2013).

CXCL9, also known as monokine induced by interferon-gamma (MIG), is a chemokine that is secreted by MSCs. It plays a crucial role in the immune response against microbial infections. CXCL9 acts by attracting immune cells, such as T cells and natural killer (NK) cells, to the site of infection. These immune cells can then eliminate the invading pathogens through various mechanisms, including the release of antimicrobial molecules and the induction of apoptosis in infected cells (Swamydas et al., 2013).

CXCL10, also known as interferon-gamma-induced protein 10 (IP-10), is another chemokine secreted by MSCs that exhibits potent anti-microbial effects. CXCL10 acts by attracting immune cells, particularly T cells, to the site of infection. Once recruited, these T cells can directly kill the invading pathogens or stimulate other immune cells to mount an effective immune response. CXCL10 also enhances the production of antimicrobial molecules, such as nitric oxide, by immune cells, further contributing to the anti-microbial activity (Swamydas et al., 2013).

CXCL11, also known as interferon-inducible T-cell alpha chemoattractant (I-TAC), is a chemokine secreted by MSCs that plays a crucial role in the immune response against microbial infections. CXCL11 attracts immune cells, including T cells and dendritic cells, to the site of infection. These immune cells can then eliminate the pathogens through various mechanisms, such as phagocytosis and the release of antimicrobial molecules. CXCL11 also promotes the activation and maturation of dendritic cells, which are essential for initiating and coordinating the immune response against microbial infections (Swamydas et al., 2013).

CCL2, also known as monocyte chemoattractant protein-1 (MCP-1), is a chemokine secreted by MSCs that exhibits anti-microbial properties. CCL2 attracts monocytes, a type of immune cell, to the site of infection. Once recruited, monocytes can differentiate into macrophages, which are highly effective in phagocytosing and eliminating microbial pathogens. CCL2 also enhances the production of antimicrobial molecules by macrophages, further contributing to the anti-microbial activity (Lee et al., 2010).

CCL5, also known as regulated upon activation, normal T cell expressed and secreted (RANTES), is a chemokine secreted by MSCs that plays a crucial role in the immune response against microbial infections. CCL5 attracts various immune cells, including T cells, eosinophils, and basophils, to the site of infection. These immune cells can then eliminate the pathogens through various mechanisms, such as the release of antimicrobial molecules and the induction of apoptosis in infected cells. CCL5 also enhances the recruitment and activation of other immune cells, further amplifying the anti-microbial response (Swamydas et al., 2013).

CXCL12, also known as stromal cell-derived factor-1 (SDF-1), is a chemokine secreted by MSCs that exhibits anti-microbial effects. CXCL12 attracts immune cells, such as T cells and monocytes, to the site of infection. Once recruited, these immune cells can directly eliminate the invading pathogens or stimulate other immune cells to mount an effective immune response. CXCL12 also enhances the production of antimicrobial molecules by immune cells, further contributing to the anti-microbial activity (Liu et al., 2018).

CX3CL1, also known as fractalkine, is a chemokine secreted by MSCs that plays a crucial role in the immune response against microbial infections. CX3CL1 attracts immune cells, including T cells and monocytes, to the site of infection. These immune cells can then eliminate the pathogens through various mechanisms, such as phagocytosis and the release of antimicrobial molecules. CX3CL1 also promotes the recruitment and activation of other immune cells, further enhancing the anti-microbial response (Swamydas et al., 2013).

The chemokines mentioned above are just a few examples of the diverse array of chemokines present in the secretome of MSCs. These chemokines work in synergy with other components of the secretome, such as growth factors and cytokines, to exert potent anti-microbial effects. Harnessing the therapeutic potential of these chemokines and the secretome as a whole holds great promise for the development of novel anti-microbial therapies. Further research is needed to fully understand the mechanisms underlying the anti-microbial effects of the secretome and to optimize its therapeutic applications.

CHAPTER EIGHT

ANTI-APOPTOTIC EFFECTS OF THE SECRETOME

Growth Factors involved in Anti-Apoptotic Effects

A poptosis, or programmed cell death, is a natural process that plays a crucial role in maintaining tissue homeostasis and eliminating damaged or unwanted cells. However, excessive or uncontrolled apoptosis can lead to tissue damage and contribute to the development of various diseases. Mesenchymal stem cells (MSCs) have been shown to possess remarkable anti-apoptotic effects, making them a promising therapeutic tool for the treatment of conditions characterized by excessive cell death (Semita et al., 2023).

The anti-apoptotic effects of MSCs are mediated by a complex interplay of various growth factors, cytokines, and chemokines present in their secretome. These bioactive molecules act in a coordinated manner to promote cell survival, inhibit apoptosis, and enhance tissue regeneration. In this section, we will explore some of the key growth factors involved in the anti-apoptotic effects of the secretome of MSCs.

Vascular Endothelial Growth Factor (VEGF) is a potent angiogenic factor that plays a crucial role in promoting the growth of new blood vessels. In addition to its angiogenic properties, VEGF has been shown to possess anti-apoptotic effects. It acts by binding to its receptors on endothelial cells and activating downstream signaling pathways that promote cell survival and inhibit apoptosis. By enhancing angiogenesis and improving blood supply to tissues, VEGF indirectly contributes to the anti-apoptotic effects of MSCs (Chun et al., 2019).

Insulin-like Growth Factor-1 (IGF-1) is a growth factor that regulates cell growth, proliferation, and survival. It has been shown to possess potent anti-apoptotic effects in various cell types. IGF-1 acts by activating the PI3K/Akt signaling pathway, which in turn inhibits apoptosis and promotes cell survival. MSCs secrete IGF-1 as part of their secretome, and this growth factor contributes to their anti-apoptotic effects (Chun et al., 2019).

Hepatocyte Growth Factor (HGF) is a multifunctional growth factor that plays a critical role in tissue repair and regeneration. It has been shown to possess potent anti-apoptotic effects in various cell types. HGF acts by binding to its receptor, c-Met, and activating downstream signaling pathways that promote cell survival and inhibit apoptosis. MSCs secrete HGF as part of their secretome, and this growth factor contributes to their anti-apoptotic effects (Chun et al., 2019).

Transforming Growth Factor-beta (TGF-beta) is a pleiotropic cytokine that regulates various cellular processes, including cell growth, differentiation, and apoptosis. TGF-beta has been shown to possess both pro-apoptotic and anti-apoptotic effects, depending on the cellular context. In the context of MSCs, TGF-beta has been found to exert anti-apoptotic effects by inhibiting apoptosis and promoting cell survival. The exact mechanisms underlying the anti-apoptotic effects of TGF-beta in MSCs are still being elucidated (Semita et al., 2023).

Platelet-Derived Growth Factor (PDGF) is a potent mitogen and chemoattractant that plays a crucial role in tissue repair and regeneration. PDGF has been shown to possess anti-apoptotic effects in various cell types. It acts by binding to its receptors on target cells and activating downstream signaling pathways that promote cell survival and inhibit apoptosis. MSCs secrete PDGF as part of their secretome, and this growth factor contributes to their anti-apoptotic effects (Chun et al., 2019).

Other Growth Factors

In addition to the growth factors mentioned above, the secretome of MSCs contains various other growth factors that contribute to their anti-apoptotic effects. These include fibroblast growth factor (FGF), nerve growth factor (NGF), and epidermal growth factor (EGF), among others. Each of these growth factors exerts its anti-apoptotic effects through specific mechanisms, such as activating survival signaling pathways or inhibiting pro-apoptotic factors (Chun et al., 2019).

In conclusion, the secretome of MSCs contains a diverse array of growth factors that contribute to their remarkable anti-apoptotic effects. These growth factors act in a coordinated manner to promote cell survival, inhibit apoptosis, and enhance tissue regeneration. Understanding the mechanisms underlying the anti-apoptotic effects of the secretome is crucial for harnessing the therapeutic potential of MSCs in the treatment of various diseases characterized by excessive cell death. Further research is needed to fully elucidate the specific roles of each growth factor and their interactions in mediating the anti-apoptotic effects of MSCs (Semita et al., 2023).

Cytokines involved in Anti-Apoptotic Effects

Apoptosis, delineated as programmed cellular demise, constitutes an intrinsic mechanism pivotal for sustaining tissue equilibrium and purging cells that are either damaged or superfluous. Nonetheless, an aberrant or hyperactive apoptotic process can precipitate tissue detriment and is implicated in the etiology of a plethora of pathologies. Mesenchymal stem cells (MSCs) have garnered attention due to their pronounced anti-apoptotic capabilities, positioning them as a potential therapeutic agent for ailments typified by exacerbated cellular mortality. These anti-apoptotic attributes are, to a significant extent, facilitated through the MSCs' secretion of a diverse spectrum of cytokines. This segment is dedicated to a detailed examination of the cytokines integral to the anti-apoptotic influence exerted by the secretome of MSCs (Jiao et al., 2021).

Interleukin-6 (IL-6)

Interleukin-6 (IL-6) is a multifunctional cytokine that plays a crucial role in regulating immune responses and inflammation. It has been shown to have potent anti-apoptotic effects in various cell types, including neurons, cardiomyocytes, and endothelial cells. IL-6 promotes cell survival by activating the Janus kinase/signal transducer and activator of transcription (JAK/STAT) signaling pathway, which leads to the upregulation of anti-apoptotic proteins such as Bcl-2 and Bcl-xL. Additionally, IL-6 can inhibit the activation of caspases, a family of proteases involved in the execution of apoptosis (Demircan et al., 2011).

Insulin-like Growth Factor-1 (IGF-1)

Insulin-like Growth Factor-1 (IGF-1) is a growth factor that plays a crucial role in promoting cell survival and proliferation. It has been shown to have potent anti-apoptotic effects in various cell types, including neurons, cardiomyocytes, and skeletal muscle cells. IGF-1 activates the phosphatidylinositol 3-kinase (PI3K)/Akt signaling pathway, which leads to the phosphorylation and inactivation of pro-apoptotic proteins such as Bad and caspase-9. Additionally, IGF-1 can upregulate the expression of anti-apoptotic proteins such as Bcl-2 and Bcl-xL (Jiao et al., 2021).

Hepatocyte Growth Factor (HGF)

Hepatocyte Growth Factor (HGF) is a growth factor that plays a crucial role in promoting tissue regeneration and repair. It has been shown to have potent anti-apoptotic effects in various cell types, including hepatocytes, cardiomyocytes, and renal tubular cells. HGF activates the PI3K/Akt signaling pathway, which leads to the phosphorylation and inactivation of pro-apoptotic proteins such as Bad and caspase-9. Additionally, HGF can upregulate the expression of anti-apoptotic proteins such as Bcl-2 and Bcl-xL (Jiao et al., 2021).

Vascular Endothelial Growth Factor (VEGF)

Vascular Endothelial Growth Factor (VEGF) is a growth factor that plays a crucial role in promoting angiogenesis, the formation of new blood vessels. It has been shown to have potent anti-apoptotic effects in various cell types, including endothelial cells and cardiomyocytes. VEGF activates the PI3K/Akt signaling pathway, which leads to the phosphorylation and inactivation of pro-apoptotic proteins such as Bad and caspase-9. Additionally, VEGF can upregulate the expression of anti-apoptotic proteins such as Bcl-2 and Bcl-xL (Jiao et al., 2021).

Transforming Growth Factor-beta (TGF-beta)

Transforming Growth Factor-beta (TGF-beta) is a multifunctional cytokine that plays a crucial role in regulating cell growth, differentiation, and apoptosis. It has been shown to have both pro-apoptotic and anti-apoptotic effects, depending on the cell type and context. In the context of MSC secretome-mediated anti-apoptotic effects, TGF-beta has been shown to promote cell survival by activating the Smad signaling pathway, which leads to the upregulation of anti-apoptotic proteins such as Bcl-2 and Bcl-xL. Additionally, TGF-beta can inhibit the activation of caspases and the release of cytochrome c, a pro-apoptotic protein (Demircan et al., 2011).

Interleukin-10 (IL-10)

Interleukin-10 (IL-10) is an anti-inflammatory cytokine that plays a crucial role in regulating immune responses and inflammation. It has been shown to have potent anti-apoptotic effects in various cell types, including immune cells and neurons. IL-10 promotes cell survival by activating the PI3K/Akt signaling pathway, which leads to the phosphorylation and inactivation of pro-apoptotic proteins such as Bad and caspase-9. Additionally, IL-10 can upregulate the expression of anti-apoptotic proteins such as Bcl-2 and Bcl-xL (Demircan et al., 2011).

Stromal-Derived Factor-1 (SDF-1)

Stromal-Derived Factor-1 (SDF-1), also known as CXCL12, is a chemokine that plays a crucial role in regulating cell migration, proliferation, and survival. It has been shown to have potent anti-apoptotic effects in various cell types, including immune cells and neurons. SDF-1 promotes cell survival by activating the PI3K/Akt signaling pathway, which leads to the phosphorylation and inactivation of pro-apoptotic proteins such as Bad and caspase-9. Additionally, SDF-1 can upregulate the expression of anti-apoptotic proteins such as Bcl-2 and Bcl-xL (Demircan et al., 2011).

In conclusion, the secretome of MSCs contains a variety of cytokines that contribute to their anti-apoptotic effects. These cytokines, including IL-6, IGF-1, HGF, VEGF, TGF-beta, IL-10, and SDF-1, promote cell survival by activating various signaling pathways and upregulating the expression of anti-apoptotic proteins. Understanding the role of these cytokines in the anti-apoptotic effects of MSCs can provide valuable insights for the development of novel therapeutic strategies for conditions characterized by excessive cell death.

Chemokines involved in Anti-Apoptotic Effects

Chemokines are a group of small proteins that play a crucial role in cell signaling and immune responses. In addition to their well-known functions in immune cell recruitment and inflammation, chemokines have also been found to have anti-apoptotic effects. These effects are particularly important in the context of regenerative medicine, as they can promote cell survival and tissue repair.

CXCL12 (SDF-1)

One of the key chemokines involved in anti-apoptotic effects is CXCL12, also known as stromal cell-derived factor 1 (SDF-1). CXCL12 has been shown to protect cells from apoptosis in various tissues and organs (He et al., 2022). It acts by binding to its receptor, CXCR4, which is expressed on the surface of many cell types including mesenchymal stem cells (MSCs). Activation of the CXCL12/CXCR4 signaling pathway leads to the activation of several downstream signaling pathways, such as the PI3K/Akt and MAPK/ERK pathways, which promote cell survival and inhibit apoptosis (He et al., 2022).

CCL2 (MCP-1)

CCL2, also known as monocyte chemoattractant protein-1 (MCP-1), is another chemokine that has been implicated in anti-apoptotic effects. CCL2 is produced by various cell types, including MSCs, and acts by binding to its receptor, CCR2. Activation of the CCL2/CCR2 signaling pathway has been shown to protect cells from apoptosis in different experimental models. The exact mechanisms by which CCL2 exerts its anti-apoptotic effects are not fully understood, but it is believed to involve the activation of survival signaling pathways and the inhibition of pro-apoptotic factors.

CX3CL1 (Fractalkine)

CX3CL1, also known as fractalkine, is a unique chemokine that exists in both membrane-bound and soluble forms. It is expressed by various cell types, including MSCs, and acts by binding to its receptor, CX3CR1. CX3CL1 has been shown to have anti-apoptotic effects in different cell types and tissues. The mechanisms underlying its anti-apoptotic effects involve the activation of survival pathways, such as the PI3K/Akt and MAPK/ERK pathways, and the inhibition of apoptotic signaling pathways.

CCL5 (RANTES)

CCL5, also known as regulated upon activation, normal T cell expressed and secreted (RANTES), is a chemokine that is produced by various cell types, including MSCs. It acts by binding to its receptors, CCR1, CCR3, and CCR5. CCL5 has been shown to have anti-apoptotic effects in different cell types and tissues. The mechanisms by which CCL5 exerts its anti-apoptotic effects involve the activation of survival signaling pathways and the inhibition of apoptotic pathways.

Other Chemokines

In addition to the chemokines mentioned above, several other chemokines have been implicated in anti-apoptotic effects. These include CCL3 (MIP-1α), CCL4 (MIP-1β), and CXCL8 (IL-8). These chemokines have been shown to protect cells from apoptosis in various experimental models. The exact mechanisms by which they exert their anti-apoptotic effects are not fully understood and may involve the activation of survival pathways and the inhibition of apoptotic signaling pathways.

Overall, the secretome of mesenchymal stem cells contains a variety of chemokines that have been shown to have anti-apoptotic effects. These chemokines act by activating survival signaling pathways and inhibiting apoptotic pathways, thereby promoting cell survival and tissue repair. Understanding the role of chemokines in the anti-apoptotic effects of the secretome is crucial for harnessing the regenerative potential of mesenchymal stem cells and developing novel therapeutic strategies for various diseases and injuries (He et al., 2022).

CHAPTER NINE

CONCLUSION AND FUTURE DIRECTIONS

T he intricate interplay between the secretome components of mesenchymal stem cells (MSCs) and their target environments encapsulates a promising frontier in regenerative medicine. As elucidated throughout this document, the secretome's multifaceted nature, encompassing an array of growth factors, cytokines, chemokines, and other bioactive molecules, orchestrates a symphony of biological processes essential for tissue repair, regeneration, and homeostasis.

Growth factors such as EGF, FGF, PDGF, and IGF, among others, have been spotlighted for their pivotal roles in stimulating cell proliferation, migration, and angiogenesis. These factors intricately contribute to the orchestration of wound healing and tissue regeneration by fostering cellular growth and differentiation, highlighting the potential of these molecules in therapeutic interventions.

Cytokines and **chemokines**, integral to the secretome, not only modulate inflammatory responses but also serve as guiding beacons for immune cells, steering them to sites of injury or inflammation. Their dual role in both promoting regenerative processes and modulating the immune response underlines their importance in maintaining tissue homeostasis and in the potential treatment of immune-related disorders.

The document further delves into the **immunomodulatory effects** of the MSC secretome, illustrating its capacity to create a conducive environment for tissue repair and regeneration. Through a complex interplay of inhibitory and stimulatory mechanisms, the secretome components help in sculpting a balanced immune landscape, mitigating inflammation, and promoting tissue integrity.

In the realm of **cell proliferation**, the secretome's influence is profound. By furnishing a biochemical milieu rich in mitogenic and chemotactic signals, it ensures the sustenance and replenishment of cellular populations critical for tissue maintenance and repair.

However, while the therapeutic potential of the MSC secretome is undeniable, it's important to acknowledge the nuances and complexities involved. The biological effects of the secretome are profoundly influenced by the specific context of its application, including the local tissue environment, the nature of the injury or disease, and the interplay with systemic factors. Hence, a deeper understanding of these contextual determinants is paramount for harnessing the full therapeutic potential of the secretome.

Future Directions

The promising landscape of mesenchymal stem cell (MSC) secretome research paves the way for transformative advancements in regenerative medicine. However, to fully harness the therapeutic potential of the secretome, several future directions need to be meticulously navigated.

Elucidation of Secretome Composition

A comprehensive understanding of the secretome's complex composition is paramount. Future studies should aim to delineate the complete array of proteins, lipids, and RNA molecules present in the MSC secretome. Advanced proteomic, lipidomic, and transcriptomic techniques will be instrumental in unveiling the intricate molecular landscape of the secretome, setting the stage for targeted therapeutic strategies.

Contextual Influence and Secretome Tailoring

Acknowledging the influence of the tissue microenvironment on secretome's efficacy is crucial. Future research should focus on how different pathological conditions affect the secretome's composition and therapeutic potential. Personalized secretome therapies, tailored to individual patient needs and specific disease contexts, could significantly enhance treatment outcomes.

Standardization and Scalability

For the MSC secretome to transition into a viable clinical therapy, establishing standardized protocols for secretome collection, processing, and storage is essential. Moreover, scalable production methods need to be developed to meet the demands of widespread clinical application, ensuring that secretome therapies are accessible and affordable.

Delivery Mechanisms

Innovative delivery methods that ensure the stability, bioactivity, and targeted delivery of secretome components are critical. Research into biomaterials and nanotechnology could provide novel platforms for the controlled release and site-specific delivery of secretome factors, enhancing their regenerative potential.

Clinical Trials and Regulatory Frameworks

Rigorous clinical trials are necessary to validate the safety and efficacy of secretome-based therapies. Developing a robust regulatory framework, encompassing quality control standards, ethical guidelines, and safety protocols, will be essential in facilitating the transition of secretome therapies from the laboratory to the clinic.

Mechanistic Insights and Predictive Models

Deepening our understanding of the mechanisms through which the secretome exerts its regenerative effects will enable the development of predictive models for treatment outcomes. Integrative approaches, combining computational modeling with experimental research, could provide valuable insights into the secretome's therapeutic mechanisms, guiding the design of optimized treatment regimens.

In conclusion, the future of MSC secretome research holds immense promise for the field of regenerative medicine. By addressing these key areas, the scientific community can unlock the full therapeutic potential of the MSC secretome, ushering in a new era of personalized, molecularly tailored treatments. As we stand on the brink of this exciting frontier, it is our collective responsibility to ensure that our scientific journey is guided by rigor, ethics, and a steadfast commitment to improving patient outcomes. The MSC secretome stands at the cusp of a new era in regenerative medicine, offering hope for the treatment of a myriad of diseases and injuries. Its complex repertoire of bioactive molecules holds the key to unlocking novel therapeutic strategies, marking a paradigm shift from conventional treatments to molecularly tailored therapies. As we continue to unravel the mysteries of the secretome, it is imperative that our explorations are underpinned by rigorous scientific inquiry, ensuring that the transition from bench to bedside is both ethically sound and scientifically robust, paving the way for a future where regenerative medicine can realize its full potential.

CHAPTER TEN

REFERENCES

Chapter 1

1. Soetjahjo, B. (2022). Mesenchymal Stem Cells Secretome and Osteoarthritis: A State of The Art. *HIP & KNEE Journal*, 3(2).

2. Bryk, M., Karnas, E., Mlost, J., Zuba-Surma, E., & Starowicz, K. (2021). Mesenchymal stem cells and extracellular vesicles for the treatment of pain: Current status and perspectives. *British Journal of Pharmacology*, 178(22), 4451-4470.

3. Smolinská, V., Boháč, M., & Danišovič, Ľ. (2023). Current status of the applications of conditioned media derived from mesenchymal stem cells for regenerative medicine. *Physiological Research*.

4. Barilani, M. (2017). Stem cell extracellular vesicles for neural regeneration.

5. Joshi, J., Dehghan Abnavi, M., & Kothapalli, C. (2019). Synthesis and secretome release by human bone marrow

mesenchymal stem cell spheroids within three-dimensional collagen hydrogels: Integrating experiments and modelling. *Journal of Tissue Engineering and Regenerative Medicine,* *13*(10), 1711-1723. https://doi.org/10.1002/term.2943

6. Deschepper, M., Paquet, J., & Petite, H. (2014). NUTRI-ENT AND O2 TENSION ARE CRUCIAL REGULA-TORS OF HUMAN MESENCHYMAL STEM CELLS (HMSCS) PARACRINE POTENTIALITIES. *Conference Proceedings.*

7. Lai, W. F., Phang, W. L. M., Govindasamy, V., Then, K., & Das, A. (2019). Understanding the multifaceted mecha-nisms of diabetic wound healing and therapeutic application of stem cells conditioned medium in the healing process. *Journal of Tissue Engineering and Regenerative Medicine,* *13*(12), 2046-2062. https://doi.org/10.1002/term.2966

8. Bellei, B., Migliano, E., & Picardo, M. (2022). Ther-apeutic potential of adipose tissue-derivatives in mod-ern dermatology. *Experimental Dermatology, 31*(1), 5-13. https://doi.org/10.1111/exd.14532

Chapter 2

1. Xu, J., Wang, D., Liu, D., Fan, Z., Zhang, H., Liu, O., Ding, G., Gao, R., Zhang, C., Ding, Y., Bromberg, J., Chen, W., Sun, L., & Wang, S. (2013). Mesenchymal stem cells: A new trend for cell therapy. Acta Pharmacologica Sinica, 34(6), 747–754.

2. Gnecchi, M., Zhang, Z., Ni, A., & Dzau, V. J. (2008). Paracrine mechanisms in adult stem cell signaling and ther-

apy. Circulation Research, 103(11), 1204–1219.

3. Uccelli, A., Moretta, L., & Pistoia, V. (2008). Mesenchymal stem cells in health and disease. Nature Reviews Immunology, 8(9), 726–736.

4. Caplan, A. I. (2011). Mesenchymal stem cells: Time to change the name! Stem Cells Translational Medicine, 1(6), 480–490.

5. Pittenger, M. F., Discher, D. E., Péault, B. M., Phinney, D. G., Hare, J. M., & Caplan, A. I. (2019). Mesenchymal stem cell perspective: cell biology to clinical progress. NPJ Regenerative Medicine, 4(1), 22.

6. Phinney, D. G., & Prockop, D. J. (2007). Concise review: Mesenchymal stem/marrow stromal cells: The state of transdifferentiation and modes of tissue repair—Current views. Stem Cells, 25(11), 2896–2902.

7. Richardson, S. M., Hoyland, J. A., Mobasheri, R., Csaki, C., Shakibaei, M., & Mobasheri, A. (2016). Mesenchymal stem cells in regenerative medicine: Opportunities and challenges for articular cartilage and intervertebral disc tissue engineering. Journal of Cellular Physiology, 231(1), 15–32.

8. Forbes, G. M., & Rosenthal, N. (2014). Mesenchymal stem cells: Mechanisms of immunomodulation and homing. Cell Transplantation, 23(6), 639–654.

9. Le Blanc, K., & Davies, L. C. (2008). Mesenchymal stromal cells and the immune system. In R. C. Sobel & D. A. Carter (Eds.), Adult Stem Cells: Biology and Methods of Analysis

(pp. 88–102). Humana Press.

10. Friedenstein, A. J., Chailakhyan, R. K., Latsinik, N. V., Panasyuk, A. F., & Keiliss-Borok, I. V. (2012). Stromal cells responsible for transferring the microenvironment of the hemopoietic tissues. Cloning in vitro and retransplantation in vivo. Transplantation, 17(4), 331–340.

11. Durum, S., Oppenheim, J., Feldmann, M., □□ □□, Vilec, J., & Nicola, N. (n.d.). *Cytokine reference: A compendium of cytokines and other mediators of host defence.* Retrieved from Google Scholar

12. Ito, S., Strickland, S., Savani, B., Jagasia, M., Melenhorst, J., Yin, F., Hensel, N., Chinian, F., Feng, X., Samsel, L., Keyvanfar, K., Hourigan, C., Muranski, P., Battiwalla, M., & Barrett, A. (2013). High levels of IL-27 occur in newly diagnosed acute myeloid leukemia (AML) and may influence outcome by suppressing T cell function. *Blood, 122*(21), 2567-2567. https://doi.org/10.1182/BLOOD.V122.21.25 67.2567

13. Ravindran, R., Khan, I., Gosselin, R., Wun, T., Krishnan, K., Luciw, P., & Janatpour, K. (2006). Multiplex assay for profiling plasma protein biomarkers for improved diagnosis and prognosis of heparin induced thrombocytopenia (HIT). *Blood, 108*(11), 4106-4106. https://doi.org/10.1182/BLO OD.V108.11.4106.4106

14. Kumar, R., Kumar, S., Singh, D., & Goel, S. (n.d.). Role of Bone Marrow Derived Autologous Mesenchymal Stem Cells in Fracture Healing in Rabbits. Retrieved from Google

Scholar

15. Wang, J., Chen, X., Yang, X., Guo, B., Li, D., Zhu, X., & Zhang, X. (2020). Positive role of calcium phosphate ceramics regulated inflammation in the osteogenic differentiation of mesenchymal stem cells. *Journal of Biomedical Materials Research Part A, 108*(4), 632–643. https://doi.org/10.100 2/jbm.a.36903

16. Adas, G., Koç, B., Adaş, M., Duruksu, G., Subaşı, C., Kemik, O., Kemik, A., Sakız, D., Kalaycı, M., Purisa, S., Unal, S., & Karaoz, E. (2016). Effects of mesenchymal stem cells and VEGF on liver regeneration following major resection. *Langenbeck's Archives of Surgery, 401*(4), 725–740. https://doi .org/10.1007/s00423-016-1380-9

17. Konala, V. B. R., Bhonde, R., & Pal, R. (2020). Secretome studies of mesenchymal stromal cells (MSCs) isolated from three tissue sources reveal subtle differences in potency. *In Vitro Cellular & Developmental Biology - Animal, 56*(9), 673–685. https://doi.org/10.1007/s11626-020-00501-1

18. Konala, V. B. R., Bhonde, R., & Pal, R. (2020). Secretome studies of mesenchymal stromal cells (MSCs) isolated from three tissue sources reveal subtle differences in potency. *In Vitro Cellular & Developmental Biology - Animal, 56*(9), 673–685. https://doi.org/10.1007/s11626-020-00501-1

19. Chernoff, G. (n.d.). The Utilization of Human Placental Mesenchymal Stem Cell Derived Exosomes in Aging Skin: An Investigational Pilot Study. https://doi.org/10.29011/ 2575-9760.001388

20. Khan, S., Barry, F., O'Brien, T., & Kerin, M. (n.d.). Advances in mesenchymal stem cell-mediated gene therapy for cancer. Retrieved from Google Scholar

21. Kim, C., Liu, R., Kucia, M., & Ratajczak, M. (2011). New Evidence That the Bioactive Lipid Ceramide-1-Phosphate (C1P) Is a Potent Chemoattractant for Mesenchymal Stromal Cells (MSC), Endothelial Progenitor Cells (EPCs) and Very Small Embryonic-Like Stem Cells (VSELs), Demonstrating Its Potential Involvement in Tissue/Organ Repair and Angiogenesis. *Blood, 118*(21), 2387. https://doi.org/1 0.1182/blood.v118.21.2387.2387

22. Konala, V. B. R., Bhonde, R., & Pal, R. (2020). Secretome studies of mesenchymal stromal cells (MSCs) isolated from three tissue sources reveal subtle differences in potency. *In Vitro Cellular & Developmental Biology - Animal, 56*(9), 673–685. https://doi.org/10.1007/s11626-020-00501-1

23. Chernoff, G. (n.d.). The Utilization of Human Placental Mesenchymal Stem Cell Derived Exosomes in Aging Skin: An Investigational Pilot Study. https://doi.org/10.29011/ 2575-9760.001388

24. Khan, S., Barry, F., O'Brien, T., & Kerin, M. (n.d.). Advances in mesenchymal stem cell-mediated gene therapy for cancer. Retrieved from Google Scholar

25. Kim, C., Liu, R., Kucia, M., & Ratajczak, M. (2011). New Evidence That the Bioactive Lipid Ceramide-1-Phosphate (C1P) Is a Potent Chemoattractant for Mesenchymal Stromal Cells (MSC), Endothelial Progenitor Cells (EPCs) and

Very Small Embryonic-Like Stem Cells (VSELs), Demonstrating Its Potential Involvement in Tissue/Organ Repair and Angiogenesis. *Blood, 118*(21), 2387. https://doi.org/1 0.1182/blood.v118.21.2387.2387

Chapter 3

1. Ullah, I., Subbarao, R. B., & Rho, G. J. (2013). Mesenchymal stem cells: A new trend for cell therapy. *Acta Biomedica, 84*(2), 134–146. Retrieved from https://www.ncbi.nlm.n ih.gov/pmc/articles/PMC3917282/

2. Lieberman, J. R., Daluiski, A., & Einhorn, T. A. (2002). The role of growth factors in the repair of bone. Biology and clinical applications. *The Journal of Bone and Joint Surgery. American Volume, 84*(6), 1032–1044. Retrieved from http s://pubmed.ncbi.nlm.nih.gov/12063331/

3. Werner, S., & Grose, R. (2003). Cytokines in wound healing. *International Journal of Molecular Medicine, 11*(3), 307–317. Retrieved from https://pubmed.ncbi.nlm.nih.g ov/12964009/

4. Zlotnik, A., & Yoshie, O. (2000). Chemokines: A new classification system and their role in immunity. *Immunity, 12*(2), 121–127. Retrieved from https://pubmed.ncbi.nlm.nih.g ov/10661407/

5. Deszcz, I. (2023). Stem Cell-Based Therapy and Cell-Free Therapy as an Alternative Approach for Cardiac Regeneration. *Stem Cells International*. https://doi.org/10.1155/2 023/2729377

6. Costela-Ruiz, V., Melguizo-Rodríguez, L., Bellotti, C., Illescas-Montes, R., Stanco, D., Arciola, C. R., & Lucarelli, E. (2022). Different Sources of Mesenchymal Stem Cells for Tissue Regeneration: A Guide to Identifying the Most Favorable One in Orthopedics and Dentistry Applications. *International Journal of Molecular Sciences, 23*(11), 6356. https://doi.org/10.3390/ijms23116356

7. Tsuji, K., Kitamura, S., & Wada, J. (2022). Mesenchymal Stem Cells-Derived Extracellular Vesicles as 'Natural' Drug Delivery System for Tissue Regeneration. *Biocell, 46*(4). https://doi.org/10.32604/biocell.2022.018594

8. Merlo, B., & Iacono, E. (2023). Beyond Canine Adipose Tissue-Derived Mesenchymal Stem/Stromal Cells Transplantation: An Update on Their Secretome Characterization and Applications. *Animals, 13*(22), 3571. https://doi.org/10.3390/ani13223571

9. Dwyer, R. M., Khan, S., Barry, F. P., O'Brien, T., & Kerin, M. J. (2010). Advances in mesenchymal stem cell-mediated gene therapy for cancer. *Stem Cell Research & Therapy, 1.* https://doi.org/10.1186/scrt25

10. Preda, M. B. (2021). Inflammation and Hypoxia Negatively Impact the Survival and Immunosuppressive Properties of Mesenchymal Stromal Cells In Vitro. *Nicolae Simionescu Institute of Cellular Biology and Pathology.* https://www.researchgate.net/profile/Mihai-Preda

11. Dwyer, R. M., Khan, S., Barry, F. P., O'Brien, T., & Kerin, M. J. (2010). Advances in mesenchymal stem cell-mediated

gene therapy for cancer. *Stem Cell Research & Therapy, 1.*
https://doi.org/10.1186/scrt25

12. Preda, M. B. (2021). Inflammation and Hypoxia Negatively
Impact the Survival and Immunosuppressive Properties of
Mesenchymal Stromal Cells In Vitro. *Nicolae Simionescu Institute of Cellular Biology and Pathology.* https://www.rese
archgate.net/profile/Mihai-Preda

13. Chernoff, G. (n.d.). The Utilization of Human Placental
Mesenchymal Stem Cell Derived Exosomes in Aging Skin:
An Investigational Pilot Study. *Journal of Clinical & Experimental Dermatology Research* https://doi.org/10.29011/2
575-9760.001388

14. Deszcz, I. (2023). Stem Cell-Based Therapy and Cell-Free
Therapy as an Alternative Approach for Cardiac Regeneration. *Stem Cells International.* https://doi.org/10.1155/2
023/2729377

15. Costela-Ruiz, V., Melguizo-Rodríguez, L., Bellotti, C.,
Illescas-Montes, R., Stanco, D., Arciola, C. R., & Lucarelli, E. (2022). Different Sources of Mesenchymal Stem Cells
for Tissue Regeneration: A Guide to Identifying the Most
Favorable One in Orthopedics and Dentistry Applications.
International Journal of Molecular Sciences, 23(11), 6356.
https://doi.org/10.3390/ijms23116356

16. Tsuji, K., Kitamura, S., & Wada, J. (2022). Mesenchymal
Stem Cells-Derived Extracellular Vesicles as 'Natural' Drug
Delivery System for Tissue Regeneration. *Biocell, 46*(4). ht
tps://doi.org/10.32604/biocell.2022.018594

17. Merlo, B., & Iacono, E. (2023). Beyond Canine Adipose Tissue-Derived Mesenchymal Stem/Stromal Cells Transplantation: An Update on Their Secretome Characterization and Applications. *Animals, 13*(22), 3571. https://doi.org/10.3390/ani13223571

18. Han, Y., Yang, J., Fang, J., Zhou, Y., Candi, E., Wang, J., Hua, D., Shao, C., & Shi, Y. (2022). The secretion profile of mesenchymal stem cells and potential applications in treating human diseases. *Signal Transduction and Targeted Therapy, 7*(1), 1-16. https://doi.org/10.1038/s41392-022-00932-0

19. Qazi, R.-e-M., Khan, I., Haneef, K., Malick, T. S., Naeem, N., Ahmad, W., Salim, A., & Mohsin, S. (2022). Combination of mesenchymal stem cells and three-dimensional collagen scaffold preserves ventricular remodeling in rat myocardial infarction model. *World Journal of Stem Cells, 14*(8), 633-649. https://doi.org/10.4252/wjsc.v14.i8.633

20. Rivera, C., Tuemmers, C., Bañados, R., Vidal-Seguel, N., & Montiel-Eulefi, E. (2020). Reduction of recurrent tendinitis scar using autologous mesenchymal stem cells derived from adipose tissue from the base of the tail in Holsteiner horses (Equus ferus caballus). *International Journal of Morphology, 38*(1), 186-191. https://doi.org/10.4067/s0717-95022020000100186

21. Zhang, J., Wu, M., Zhang, X., Yang, M., Xiong, T.-L., & Zhi, W. (2023). Adult bone marrow-derived mesenchymal stem cells seeded on tissue-engineered cardiac patch contribute to myocardial scar remodeling and enhance revascularization

in a rabbit model of chronic myocardial infarction. *Heart Surgery Forum, 26*(1), E087-E095. https://doi.org/10.153 2/hsf.5067

Chapter 4

1. Xu, J., Wang, W., Ludeman, M., Cheng, K., Hayami, T., Lotz, J. C., & Kapila, S. (2013). Mesenchymal stem cells: a new trend for cell therapy. *Acta Pharmacologica Sinica, 34*(6), 747–754. https://doi.org/10.1038/aps.2013.73

2. Elahi, K. C., Klein, G., Avci-Adali, M., Sievert, K. D., Mac-Neil, S., & Aicher, W. K. (2019). Mesenchymal stem cells: Cell therapy and regeneration potential. *Journal of Tissue Engineering and Regenerative Medicine, 13*(9), 1749–1760. https://doi.org/10.1002/term.2914

3. Teixeira, F. G., Carvalho, M. M., Sousa, N., & Salgado, A. J. (2013). The secretome of mesenchymal stem cells: potential implications for neuroregeneration. *Biochimie, 95*(12), 2246–2256. https://doi.org/10.1016/j.biochi.2013.06.020

4. Tofiño-Vian, M., Guillén, M. I., Pérez Del Caz, M. D., Silvestre, A., & Alcaraz, M. J. (2018). Mesenchymal stem cell-derived exosomes: a new therapeutic approach to osteoarthritis? *Arthritis Research & Therapy, 20*(1), 18. https://doi.org/10.1186/s13075-017-1514-7

5. Börger, V., Bremer, M., Ferrer-Tur, R., Gockeln, L., Stambouli, O., Becic, A., & Giebel, B. (2017). Mesenchymal stem/stromal cell-derived exosomes and their potential as novel immunomodulatory therapeutic agents. *International Journal of Molecular Sciences, 18*(7), 1450. https://doi.org/

10.3390/ijms18071450

6. Chen, X., Chen,Chen, X., Chen, Y.-H., Hou, Y., Song, P., Zhou, M., Nie, M., & Liu, X. (2019). Modulation of proliferation and differentiation of gingiva-derived mesenchymal stem cells by concentrated growth factors: Potential implications in tissue engineering for dental regeneration and repair. *International Journal of Molecular Medicine, 44*(1), 17-30. https://doi.org/10.3892/ijmm.2019.4172

7. Hade, M. D., Suire, C. N., & Suo, Z. (2021). Mesenchymal Stem Cell-Derived Exosomes: Applications in Regenerative Medicine. *Cells, 10*(8), 1959. https://doi.org/10.3390/cells 10081959

8. Hormozi, A., Hasanzadeh, S., Ebrahimi, F., Daei, N., Hajimortezayi, Z., Mehdizadeh, A., & Zamani, M. (2023). Treatment with Exosomes Derived from Mesenchymal Stem Cells: A New Window of Healing Science in Regenerative Medicine. *Current Stem Cell Research & Therapy*. https://doi.org/10.2174/1574888X18666230824165014

9. Jiang, L., Lu, J., Chen, Y., Lyu, K., Long, L., Wang, X., Liu, T., & Li, S. (2023). Mesenchymal stem cells: An efficient cell therapy for tendon repair (Review). *International Journal of Molecular Medicine, 51*(6). https://doi.org/10.3892/ijmm .2023.5273

10. Kwaan, H., & Lindholm, P. (2019). Emergent Paradigms of Thrombosis and Cancer (Part II): More on Thrombosis and Cancer. *Seminars in Thrombosis and Hemostasis, 45*(6), 577-582. https://doi.org/10.1055/s-0039-1694764

11. Ozaki, Y., Nishimura, M., Sekiya, K., Suehiro, F., Kanawa, M., Nikawa, H., Hamada, T., & Kato, Y. (2007). Comprehensive analysis of chemotactic factors for bone marrow mesenchymal stem cells. *Stem Cells and Development, 16*(1), 119-129. https://doi.org/10.1089/SCD.2006.0032

12. Wang, J., Chen, X., Yang, X., Guo, B., Li, D., Zhu, X., & Zhang, X. (2020). Positive role of calcium phosphate ceramics regulated inflammation in the osteogenic differentiation of mesenchymal stem cells. *Journal of Biomedical Materials Research Part A, 108*(3), 632-643. https://doi.org/10.1002/jbm.a.36903

13. Khan, S., Barry, F., O'Brien, T., & Kerin, M. (n.d.). Advances in mesenchymal stem cell-mediated gene therapy for cancer. [Link not available]

Chapter 5

1. Caplan, A. I. (2017). Mesenchymal stem cells: environment responsive therapeutics for regenerative medicine. *Experimental & Molecular Medicine, 49*(11), e386. https://www.nature.com/articles/emm2017148

2. Uccelli, A., Moretta, L., & Pistoia, V. (2008). Mesenchymal stem cells: mechanisms of immunomodulation and homing. *Cell Stem Cell, 2*(6), 602-609. https://www.cell.com/cell-stem-cell/fulltext/S1934-5909(08)00251-6

3. Maggini, J., Mirkin, G., Bognanni, I., Holmberg, J., Piazzón, I. M., Nepomnaschy, I., ... & Rabinovich, G. A. (2010). Immunomodulatory properties of mesenchymal stem cells:

cytokines and factors. *American Journal of Reproductive Immunology, 63*(6), 467-476. https://onlinelibrary.wiley.com/doi/abs/10.1111/j.1600-0897.2009.00786.x

4. Ullah, I., Subbarao, R. B., & Rho, G. J. (2013). Mesenchymal stem cells: a new trend for cell therapy. *Acta Biomedica, 84*(2), 134-148. https://www.mattioli1885journals.com/index.php/actabiomedica/article/view/2679

5. Teixeira, F. G., Carvalho, M. M., Sousa, N., & Salgado, A. J. (2013). The secretome of mesenchymal stem cells: potential implications for neuroregeneration. *Biochimie, 95*(12), 2246-2256. https://www.sciencedirect.com/science/article/abs/pii/S0300908413001844

6. Esteban-Blanco, M., González-Fernández, M., & Villar-Suárez, V. (n.d.). Protective Effect of Mesenchymal Stem Cells Derived Secretome in an in Vitro Pro-Inflammatory Model of Intervertebral Discogenic Pain.

7. Lunyak, V., Amaro-Ortiz, A., & Gaur, M. (2017). Mesenchymal Stem Cells Secretory Responses: Senescence Messaging Secretome and Immunomodulation Perspective. *Frontiers in Genetics, 8*, 220. https://doi.org/10.3389/fgene.2017.00220

8. Soetjahjo, B. (2022). Mesenchymal Stem Cells Secretome and Osteoarthritis: A State of The Art. *Hip & Knee Journal, 3*(2). https://dx.doi.org/10.46355/hipknee.v3i2.133

9. Teixeira, F., Carvalho, M., Sousa, N., & Salgado, A. (2013). Mesenchymal stem cells secretome: a new paradigm for cen-

tral nervous system regeneration? *Cellular and Molecular Life Sciences, 70*(20), 3871-3882. https://doi.org/10.1007/s00018-013-1290-8

10. Cao X. et al. (2017). Effect and safety evaluation of 50-Hz 1-mT pulsed electromagnetic field on the immunomodulation of human umbilical cord mesenchymal stem cells. [Link not available]

11. Pinheiro A. de O. et al. (2020). Characterization and Immunomodulation of Canine Amniotic Membrane Stem Cells. Stem Cells and Cloning: Advances and Applications. https://doi.org/10.2147/SCCAA.S237686

12. Qian D. et al. (2015). Bone Marrow-Derived Mesenchymal Stem Cells Repair Necrotic Pancreatic Tissue and Promote Angiogenesis by Secreting Cellular Growth Factors Involved in the SDF-1α/CXCR4 Axis in Rats. Stem Cells International. https://doi.org/10.1155/2015/306836

Chapter 6

1. Xu, J., Wang, D., Liu, D., Fan, Z., Zhang, H., Liu, O., Ding, G., Gao, R., Zhang, C., Ding, Y., Bromberg, J., Chen, W., Sun, L., & Wang, S. (2013). Mesenchymal stem cells: A new trend for cell therapy. *Acta Pharmacologica Sinica*, 34(6), 747–754. https://doi.org/10.1038/aps.2013.38

2. Uccelli, A., Moretta, L., & Pistoia, V. (2008). Mesenchymal stem cells in health and disease. *Nature Reviews Rheumatology*, 4(9), 481–489. https://doi.org/10.1038/nrrheum.2009.191

3. Klopp, A. H., Gupta, A., Spaeth, E., Andreeff, M., & Marini, F. (2011). Concise review: Dissecting a discrepancy in the literature: Do mesenchymal stem cells support or suppress tumor growth? *Stem Cells*, 29(1), 11–19. https://doi.org/1 0.1002/stem.559

4. Le Blanc, K., & Mougiakakos, D. (2012). Mesenchymal stem cells and immunomodulation: Current status and future prospects. *Cytotherapy*, 14(2), 213–215. https://doi.org/1 0.3109/14653249.2011.652793

5. Caplan, A. I. (2009). Why are MSCs therapeutic? New data: New insight. *Journal of Pathology*, 217(2), 318–324. https: //doi.org/10.1002/path.2469

6. Adas, G., Çukurova, Z., Yasar, K. K., Yilmaz, R., Isiksacan, N., Kasapoğlu, P., ... & Karaoz, E. (2021). The Systematic Effect of Mesenchymal Stem Cell Therapy in Critical COVID-19 Patients: A Prospective Double Controlled Trial. *Cell Transplantation*, 30, 09636897211024942. https:/ /dx.doi.org/10.1177/09636897211024942

7. Jafri, J., Saini, D., Chaudhary, P., Maurya, A., Verma, G. K., Gupta, A. K., ... & Mirza-Shariff, A. A. (2022). Exploring the Immunomodulatory Aspect of Mesenchymal Stem Cells for Treatment of Severe Coronavirus Disease 19. *Cells*, 11(14), 2175. https://dx.doi.org/10.3390/cells11142175

8. Ziegler, C. G., Van Sloun, R., Gonzalez, S., Whitney, K. E., DePhillipo, N. N., Kennedy, M., ... & LaPrade, R. (2019). Characterization of Growth Factors, Cytokines and Chemokines in Bone Marrow Concentrate and Platelet

Rich Plasma: A Prospective Analysis. *Orthopaedic Journal of Sports Medicine, 7*(7_suppl), 2325967119S00284. https://dx.doi.org/10.1177/2325967119S00284

9. Chaudhary, J., Saini, D., Chaudhary, P., Maurya, A., Verma, G. K., Gupta, A. K., ... & Mirza-Shariff, A. A. (2022). Exploring the Immunomodulatory Aspect of Mesenchymal Stem Cells for Treatment of Severe Coronavirus Disease 19. *Cells, 11*(14), 2175. https://dx.doi.org/10.3390/cells11142 175

10. Jiao, Z., Ma, Y., Zhang, Q., Wang, Y., Liu, T., Liu, X., ... & Wang, H. (2021). The adipose-derived mesenchymal stem cell secretome promotes hepatic regeneration in miniature pigs after liver ischaemia-reperfusion combined with partial resection. *Stem Cell Research & Therapy, 12*(1), 284. https ://dx.doi.org/10.1186/s13287-021-02284-y

11. Raj, A., Kheur, S., Khurshid, Z., Sayed, M., Mugri, M., Almasri, M. A., ... & Patil, S. (2021). The Growth Factors and Cytokines of Dental Pulp Mesenchymal Stem Cell Secretome May Potentially Aid in Oral Cancer Proliferation. *Molecules, 26*(18), 5683. https://dx.doi.org/10.3390/mole cules26185683

12. Shandil, R., Dhup, S., & Narayanan, S. (2022). Evaluation of the Therapeutic Potential of Mesenchymal Stem Cells (MSCs) in Preclinical Models of Autoimmune Diseases. *Stem Cells International, 2022*, Article ID 6379161. https://dx.doi.org/10.1155/2022/6379161

13. Abdul-Hamid, M., Ahmed, R. H., Ahmed, R. R., Galaly,

S. R., Moustafa, N., Abourehab, M. A. S., Abdelgawad, M. A., & Ahmed, O. M. (2023). Mesenchymal Stem Cells and Curcumin Effectively Mitigate Freund's Adjuvant-induced Arthritis via their Anti-inflammatory and Gene Expression of COX-1, IL-6 and IL-4. *Current Pharmaceutical Design.* https://dx.doi.org/10.2174/187153032366623 0223143011

14. Mani, A., Hotra, J. W., Blackwell, S. C., Goetzl, L., & Refuerzo, J. (2023). Mesenchymal stem cells suppress inflammatory cytokines in lipopolysaccharide exposed preterm and term human pregnant myometrial cells. *American Journal of Perinatology.* https://dx.doi.org/10.1055/a-2216-9194

15. Karimi, M., Maghsoud, Z., & Halabian, R. (2021). Effect of Preconditioned Mesenchymal Stem Cells with Nisin Prebiotic on the Expression of Wound Healing Factors Such as TGF-β21, FGF-2, IL-1, IL-6, and IL-10. *Stem Cell Reviews and Reports.* https://dx.doi.org/10.1007/s40883-021-0019 4-2

16. Yi, P., Xu, X., Qiu, B., & Li, H. (2020). Impact of chitosan membrane culture on the expression of pro- and anti-inflammatory cytokines in mesenchymal stem cells. *Experimental and Therapeutic Medicine.* https://dx.doi.org/10.3 892/etm.2020.9108

17. Arezoo Gowhari Shabgah, A., Mostafaie, A., Yari, K., & Shahneh, F. Z. (2018). Possible Anti-inflammatory Effects of Mesenchymal Stem Cells Transplantation via Changes in CXCL8 Levels in Patients with Refractory Rheumatoid

Arthritis. *Iranian Journal of Medical and Cellular Micro-biology, 8*(3), 191-197. https://dx.doi.org/10.22088/IJMCM.BUMS.8.3.191

18. Panahipour, L., Stähli, A., & Haiden, N. (2022). Blocking of Caspases Exerts Anti-Inflammatory Effects on Periodontal Cells. *Life, 12*(7), 1045. https://dx.doi.org/10.3390/life12071045

19. Germano, G., Frapolli, R., Belgiovine, C., Anselmo, A., Pesce, S., Liguori, M., ... & Mantovani, A. (2010). Antitumor and anti-inflammatory effects of trabectedin on human myxoid liposarcoma cells. *Cancer Research, 70*(6), 2235-2244. https://dx.doi.org/10.1158/0008-5472.CAN-09-2335

20. Darwish, I., Banner, D., Mubareka, S., Kim, H., Besla, R., Kelvin, D. J., ... & Liles, W. C. (2013). Mesenchymal Stromal (Stem) Cell Therapy Fails to Improve Outcomes in Experimental Severe Influenza. *PLOS ONE, 8*(8), e71761. https://dx.doi.org/10.1371/journal.pone.0071761

Chapter 7

1. Egea, V., Regenfuss, B., Luttun, A., & Carmeliet, P. (2022). Properties and fate of human mesenchymal stem cells upon miRNA let-7f-promoted recruitment to atherosclerotic plaques. *Cardiovascular Research.* https://doi.org/10.1093/cvr/cvac022

2. Widowati, W., Wijaya, L., Murti, H., Fauziah, N., Maesaroh, M., & Fauzi, A. (2021). Potential of Human Wharton's

Jelly Mesenchymal Stem Cells (hWJMSCs) Secretome for COVID-19 Adjuvant Therapy Candidate. In *2021 International Conference on Health Informatics, Medical, Biological Engineering, and Sciences (InHeNce)*. IEEE. https://doi.org/10.1109/InHeNce52833.2021.9537290

3. Kaur, K. (2020). Mesenchymal stem cells novel therapeutic option besides their stem cell properties utilizing the niche effect. *ResearchGate*. Retrieved from https://www.researchgate.net/publication/342344622_Mesenchymal_stem_cells_novel_therapeutic_option_besides_their_stem_cell_properties_utilizing_the_niche_effect

4. Cortes-Araya, Y., Amilon, K., Rink, B. E., Black, G., Lisowski, Z., Donadeu, F. X., & Esteves, C. L. (2018). Comparison of Antibacterial and Immunological Properties of Mesenchymal Stem/Stromal Cells from Equine Bone Marrow, Endometrium, and Adipose Tissue. *Stem Cells and Development, 27*(21), 1518-1525. https://doi.org/10.1089/scd.2017.0241

5. Chung, W. (2013). Therapeutic potentials of antimicrobial peptides. *Journal of Bioanalysis & Biomedicine, 5*(2). https://dx.doi.org/10.4172/1948-593X.1000078

6. Ghosh, S. K., & Weinberg, A. (2021). Ramping Up Antimicrobial Peptides Against Severe Acute Respiratory Syndrome Coronavirus-2. *Frontiers in Molecular Biosciences, 8*. https://dx.doi.org/10.3389/fmolb.2021.671263

7. Krasnodembskaya, A., Song, Y., Fang, X., Gupta, N., Serikov, V., Lee, J. W., & Matthay, M. A. (2010). Antibac-

terial effect of human mesenchymal stem cells is mediated in part from secretion of the antimicrobial peptide LL-37. *Stem Cells, 28*(12), 2229-2238. https://dx.doi.org/10.1002/stem .544

8. Silva-Carvalho, A. É., Pinto, F. J., & Sarmento, B. (2021). Dissecting the relationship between antimicrobial peptides and mesenchymal stem cells. *Pharmacology & Therapeutics, 227*. https://dx.doi.org/10.1016/j.pharmthera.2021.1 07868

9. Tecle, T., Tripathi, S., & Hartshorn, K. L. (2010). Review: Defensins and cathelicidins in lung immunity. *Innate Immunity, 16*(3), 151-159. https://dx.doi.org/10.1177/1753 425909354621

10. Asami, T., Ishii, M., Fujii, H., Namkoong, H., Tasaka, S., Matsushita, K., ... & Hasegawa, N. (2013). Anti-inflammatory roles of mesenchymal stem cells in lung injury induced by the influenza virus. *Influenza Research and Treatment, 2013*. Link

11. Brandau, S., Jakob, M., Bruderek, K., Bootz, F., Giebel, B., Radtke, S., ... & Lang, S. (2014). Mesenchymal Stem Cells Augment the Anti-Bacterial Activity of Neutrophil Granulocytes. *PLoS ONE, 9*. Link

12. Konala, V. B. R., Mamidi, M. K., Bhonde, R., Das, A. K., Pochampally, R., & Pal, R. (2016). The current landscape of the mesenchymal stromal cell secretome: A new paradigm for cell-free regeneration. *Cytotherapy, 18*(1), 13-24. Link

13. Kyurkchiev, D., Bochev, I., Ivanova-Todorova, E., Mourd-jeva, M., Oreshkova, T., Belemezova, K., & Kyurkchiev, S. (2014). Secretion of immunoregulatory cytokines by mesenchymal stem cells. *World Journal of Stem Cells, 6*(5), 552-570. Link

14. Lee, J. W., Krasnodembskaya, A., McKenna, D. H., Song, Y., Abbott, J., & Matthay, M. A. (2013). Therapeutic effects of human mesenchymal stem cells in ex vivo human lungs injured with live bacteria. *American Journal of Respiratory and Critical Care Medicine, 187*(7), 751-760. Link

15. Madrigal, M., Rao, K. S., & Riordan, N. H. (2014). A review of therapeutic effects of mesenchymal stem cell secretions and induction of secretory modification by different culture methods. *Journal of Translational Medicine, 12.* Link

16. Naserian, S., Shamdani, S., Arouche, N., & Uzan, G. (2020). Regulatory T cell induction by mesenchymal stem cells depends on the expression of TNFR2 by T cells. *Stem Cell Research & Therapy, 11.* Link

17. Prockop, D. J., & Oh, J. Y. (2012). Mesenchymal Stem/Stromal Cells (MSCs): Role as Guardians of Inflammation. *Molecular Therapy: The Journal of the American Society of Gene Therapy, 20*(1), 14-20. Link

18. Ranganath, S. H., Levy, O., Inamdar, M. S., & Karp, J. M. (2012). Harnessing the mesenchymal stem cell secretome for the treatment of cardiovascular disease. *Cell Stem Cell, 10*(3), 244-258. Link

19. Shigdar, S., Li, Y., Bhattacharya, S., O'Connor, M., Pu, C., Lin, J., ... & Duan, W. (2014). Inflammation and cancer stem cells. *Cancer Letters, 345*(2), 271-278. Link

20. Weiss, A. R. R., & Dahlke, M. H. (2019). Immunomodulation by Mesenchymal Stem Cells (MSCs): Mechanisms of Action of Living, Apoptotic, and Dead MSCs. *Frontiers in Immunology, 10.* Link

21. Lee, M. J., Kim, J., Kim, M. Y., Bae, Y. S., Ryu, S. H., Lee, T. G., & Kim, J. H. (2010). Proteomic analysis of tumor necrosis factor-alpha-induced secretome of human adipose tissue-derived mesenchymal stem cells. *Journal of Proteome Research, 9*(4), 1754–1762. Link

22. Liu, L., Chen, J., Zhang, X., Sun, Q., Yang, L., Liu, A., ... & Qiu, H. (2018). Chemokine receptor 7 overexpression promotes mesenchymal stem cell migration and proliferation via secreting Chemokine ligand 12. *Scientific Reports, 8*(1), 204. Link

23. Swamydas, M., Ricci, K., Rego, S. L., & Dréau, D. (2013). Mesenchymal stem cell-derived CCL-9 and CCL-5 promote mammary tumor cell invasion and the activation of matrix metalloproteinases. *Cell Adhesion & Migration, 7*(3), 315–324. Link

Chapter 8

1. Xu, Y., & Zhang, L. (2017). Mechanism of tumor tolerance mediated by immune microenvironment. Journal of International Oncology, 44, 594-596. DOI: 10.3760/CMA.J.IS

SN.1673-422X.2017.08.009

2. Rosca, A.-M., Rayia, D. M. A., & Tutuianu, R. (2016). Emerging Role of Stem Cells - Derived Exosomes as Valuable Tools for Cardiovascular Therapy. Current Stem Cell Research & Therapy, 12(2), 134-138. DOI: 10.2174/1574 888X10666151026115320

3. Strioga, M., Viswanathan, S., Darinskas, A., Slabý, O., & Michálek, J. (2012). Same or not the same? Comparison of adipose tissue-derived versus bone marrow-derived mesenchymal stem and stromal cells. Stem Cells and Development, 21(14), 2724-2752. DOI: 10.1089/scd.2011.0722

4. Han, Y.-X., Wang, Y., Xu, Z., Li, J., Yang, J., Li, Y., Shang, Y.-t., & Luo, J. (2013). Effect of bone marrow mesenchymal stem cells from blastic phase chronic myelogenous leukemia on the growth and apoptosis of leukemia cells. Oncology Reports, 30(2), 1007-1013. DOI: 10.3892/or.2013.2518

5. Nagaishi, K., Mizue, Y., Chikenji, T. S., Otani, M., Nakano, M., Konari, N., & Fujimiya, M. (2016). Mesenchymal stem cell therapy ameliorates diabetic nephropathy via the paracrine effect of renal trophic factors including exosomes. Scientific Reports, 6. DOI: 10.1038/srep34842

6. Wang, W., & Zeng, C. (2012). EFFECTS OF PHD2 ON PARACRINE EFFECT OF ADIPOSE DERIVED MESENCHYMAL STEM CELLS MEDIATED CARDIO-PROTECTION. Heart, 98, E69 - E69. DOI: 10.1136/he artjnl-2012-302920a.172

7. Xu, H., Zhou, Y., Li, W., Zhang, B., Zhang, H., Zhao, S. -l., Zheng, P., Wu, H.-y., & Yang, J. (2018). Tumor-derived mesenchymal-stem-cell-secreted IL-6 enhances resistance to cisplatin via the STAT3 pathway in breast cancer. Oncology Letters, 15, 9142 - 9150. DOI: 10.3892/ol.2018.8463

8. Ye, Q., Xu, H., Liu, S., Li, Z., Zhou, J., Ding, F., Zhang, X., Wang, Y., Jin, Y., & Wang, Q. (2022). Apoptotic extra-cellular vesicles alleviate Pg-LPS induced inflammation of macrophages via AMPK/SIRT1/NF-κB pathway and in-hibit adjacent osteoclasts formation. Journal of Periodon-tology. DOI: 10.1002/JPER.21-0657

9. Li, B., Leung, J. Y., Chan, L. Y., Yiu, W., Li, Y., Lok, S. W. Y., Liu, W. H., Chan, K. W., Tse, H., Lai, K., & Tang, S. (2019). Amelioration of Endoplasmic Reticulum Stress by Mes-enchymal Stem Cells via Hepatocyte Growth Factor/c-Met Signaling in Obesity-Associated Kidney Injury. Stem Cells Translational Medicine, 8, 898 - 910. DOI: 10.1002/sctm. 18-0265

10. Semita, N., Novembri, D., Utomo, H., Suroto, & Surgeon, M. D. (2023). Mechanism of spinal cord injury regeneration and the effect of human neural stem cells-secretome treat-ment in rat model. *World Journal of Orthopedics, 14*(2), 64. https://dx.doi.org/10.5312/wjo.v14.i2.64

11. Chun, S., Lim, J.-O., Lee, E. H., Han, M.-H., Ha, Y., Lee, J. N., ... Kwon, T. (2019). Preparation and Characterization of Human Adipose Tissue-Derived Extracellular Matrix, Growth Factors, and Stem Cells: A Concise Review. *Tissue*

Engineering and Regenerative Medicine, 16(4), 385–393. h ttps://dx.doi.org/10.1007/s13770-019-00199-7

12. He, X., Li, C., Yin, H., Tan, X., Yi, J., Tian, S., Wang, Y., & Liu, J. (2022). Mesenchymal stem cells inhibited the apop-tosis of alveolar epithelial cells caused by ARDS through CXCL12/CXCR4 axis. [Journal Name], [Volume(Issue)], [Page Numbers]. https://dx.doi.org/10.1080/21655979.2 022.2052652

CHAPTER ELEVEN

GLOSSARY

1. **Mesenchymal Stem Cells (MSCs)**: A type of multipotent stem cell that can differentiate into a variety of cell types, including osteoblasts (bone cells), chondrocytes (cartilage cells), and adipocytes (fat cells).

2. **Immunomodulation**: The process of modifying the immune response or the functioning of the immune system, either by enhancing or suppressing it.

3. **Cell Therapy**: A therapy in which cellular material is injected, grafted, or implanted into a patient; typically involves the transfer of intact, live cells.

4. **Osteoblasts**: Cells with a single nucleus that synthesize bone.

5. **Chondrocytes**: The cells responsible for cartilage formation.

6. **Adipocytes**: Cells that store fat and are found in adipose

tissue.

7. **Macrophage**: A type of white blood cell that engulfs and digests cellular debris, foreign substances, microbes, and cancer cells in a process called phagocytosis.

8. **Homing**: The process by which cells migrate from one area of the body to another, where they are needed for repair or defense.

9. **Cell Differentiation**: The process by which a cell changes from one cell type to another, typically becoming more specialized.

10. **Phagocytosis**: The process by which a cell uses its plasma membrane to engulf a large particle, forming an internal compartment known as a phagosome.

11. **Immunosuppression**: The reduction of the activation or efficacy of the immune system.

12. **Transplantation**: The process of transferring cells, tissues, or organs from one site to another, either within the same person or from one individual to another.

13. **Inflammation**: A biological response to harmful stimuli, such as pathogens, damaged cells, or irritants, and is a protective response involving immune cells, blood vessels, and molecular mediators.

14. **Cytokines**: Small proteins that are important in cell signaling. They are released by cells and affect the behavior of other cells, and sometimes the releasing cell itself.

16. **Autoimmune Diseases**: Diseases in which the body's immune system attacks healthy cells.

16. **Regenerative Medicine**: A branch of medicine that develops methods to regrow, repair, or replace damaged or diseased cells, organs, or tissues.

17. **Tissue Engineering**: An interdisciplinary field that applies the principles of biology and engineering to the development of functional substitutes for damaged tissue.

18. **Stem Cells**: Cells with the potential to develop into many different types of cells in the body. They serve as a repair system for the body.

19. **Multipotent**: The ability of stem cells to develop into multiple, but limited, cell types in the body.

20. **Extracellular Matrix (ECM)**: A collection of extracellular molecules secreted by cells that provides structural and biochemical support to the surrounding cells.

21. **Angiogenesis**: The formation of new blood vessels, a process critical for wound healing and the formation of granulation tissue.

22. **Apoptosis**: The process of programmed cell death that occurs in multicellular organisms.

23. **Biocompatibility**: The ability of a material to perform with an appropriate host response in a specific application.

24. **Chronic Disease**: A disease that is long-lasting or recurrent.

Chronic diseases are characterized by long-term effects and slow progression.

25. **Clinical Trials**: Research studies performed in people that are aimed at evaluating a medical, surgical, or behavioral intervention.

26. **Cytotoxicity**: The quality of being toxic to cells. Examples include causing cell death or inhibiting cell growth.

27. **Differentiation**: The process by which a less specialized cell becomes a more specialized cell type.

28. **Gene Therapy**: A technique that uses genes to treat or prevent disease.

29. **Hematopoietic Stem Cells (HSCs)**: Stem cells that give rise to all the other blood cells through the process of hematopoiesis.

30. **Immunotherapy**: A type of cancer treatment that helps your immune system fight cancer.

31. **In Vitro**: Studies or processes performed or taking place in a test tube, culture dish, or elsewhere outside a living organism.

32. **In Vivo**: Studies or processes performed or taking place in a living organism.

33. **Inflammation**: A localized physical condition in which part of the body becomes reddened, swollen, hot, and often painful, especially as a reaction to injury or infection.

34. **Lymphocytes**: A type of white blood cell that is part of the immune system. There are two main types: B cells and T cells.

35. **Malignant**: Refers to cancer cells that can invade and kill nearby tissue and spread to other parts of the body.

36. **Chemokines**: Chemokines are a family of small cytokines, or signaling proteins secreted by cells. Their name is derived from their ability to induce directed chemotaxis in nearby responsive cells; they are chemotactic cytokines. Chemokines are primarily involved in the migration of cells, especially white blood cells (leukocytes), towards sites of inflammation or infection in the body. They play a critical role in the immune system's response to pathogens and injury.

37. **Cytokines**: Cytokines are a broad and loose category of small proteins (~5–20 kDa) that are important in cell signaling. They are released by cells and affect the behavior of other cells, and sometimes the releasing cell itself. Cytokines include chemokines, interferons, interleukins, lymphokines, and tumor necrosis factors but are not limited to these categories. They are crucial in fighting off infections and in mediating and regulating immunity, inflammation, and hematopoiesis (the formation of blood cellular components).

38. **Interleukins**: Interleukins are a group of cytokines that were first seen to be expressed by white blood cells (leukocytes). The term interleukin derives from (inter-) "as a means of communication", and (-leukin) deriving from leukocytes,

white blood cells that use interleukins to communicate. They are a subset of a larger group of cellular messenger molecules called cytokines. Interleukins play a significant role in the immune system by regulating cell growth, differentiation, and motility. They are particularly important in stimulating immune responses, such as inflammation.

39. **Growth Factor:** A biologically active protein that stimulates cellular growth, proliferation, healing, and differentiation. Growth factors function as signaling molecules, binding to specific receptors on target cells, thereby regulating various cellular processes. These substances are integral to tissue maintenance, repair, and the regulation of cellular life cycles in multicellular organisms. As a subset of cytokines, growth factors are pivotal in cell signaling and are particularly significant in the context of tissue regeneration and repair facilitated by mesenchymal stem cells in regenerative medicine.

www.ingramcontent.com/pod-product-compliance
Lightning Source LLC
Chambersburg PA
CBHW071048290526
45795CB00004B/1375